Characterization, Design, and Processing of Nanosize Powders and Nanostructured Materials

Characterization, Design, and Processing of Nanosize Powders and Nanostructured Materials

Ceramic Transactions Series, Volume 190

Proceedings of the 6th Pacific Rim Conference on Ceramic and Glass Technology (PacRim6); September 11–16, 2005; Maui, Hawaii

Edited by
Kevin G. Ewsuk
Yury Gogotsi

The American Ceramic Society

WILEY-INTERSCIENCE

A JOHN WILEY & SONS, INC., PUBLICATION

Published by John Wiley & Sons, Inc., Hoboken, New Jersey
Published simultaneously in Canada.

For general information on our other products and services please contact our Customer Care Department within the U.S. at 877-762-2974, outside the U.S. at 317-572-3993 or fax 317-572-4002.

Wiley also publishes its books in a variety of electronic formats. Some content that appears in print, however, may not be available in electronic format.

Library of Congress Cataloging-in-Publication Data is available.

ISBN-13 978-0-470-08033-7
ISBN-10 0-470-08033-7

10 9 8 7 6 5 4 3 2 1

Contents

Forming

Sintering and Properties

Nanocomposites and Nanostructures

Preface

Characterization, design, and processing are critical in synthesizing high-quality nano-scale powders and structures, to reproducible fabricate high-performance ceramics with nanoscale structures, and to develop multifunctional ceramic materials and structures. The "Characterization and Processing of Nanosize Powders and Particles" symposium and the "Nanoscale and Multifunctional Materials" symposium at PacRim6, The 6th Pacific Rim Conference on Ceramic and Glass Technology, held in Kapalua, Hawaii USA in September 2005, addressed these topics. The former featured: 1) techniques to characterize nanosize powders and nanoparticle dispersions; 2) green processing of nanopowders; and 3) sintering and microstructure of nanoparticle assemblies. The latter addressed design rules and utilizing compositional and morphological arrangements to produce nanostructured and multifunctional materials. This conference proceedings represents a compilation of experimental and theoretical studies from the two symposia that discuss some of the latest scientific and technological developments in nanopowders, nanostructured ceramics, and multifunctional ceramic materials.

KEVIN G. EWSUK
Sandia National Laboratories

YURY GOGOTSI
Drexel University

Synthesis

SYNTHESIS OF HIGH PURITY β–SiAlON NANOPOWDER FROM A ZEOLITE BY GAS-REDUCTION-NITRIDATION

Tomohiro Yamakawa, Toru Wakihara, Junichi Tatami, Katsutoshi Komeya and Takeshi Meguro
Graduate School of Environment and Information Sciences, Yokohama National University
79-7, Tokiwadai, Hodogaya-ku
Yokohama, 240-8501, Japan

ABSTRACT

β-SiAlON powders were synthesized by gas-reduction-nitridation of zeolite using NH_3 and C_3H_8 as reactant gases. Zeolite changed from crystalline into glassy phase around 900 °C. The peak intensity of β-SiAlON increased with increasing firing temperature and soaking time. It was confirmed that products were high purity β-SiAlON. Specific surface areas of the products nitrided at 1400 °C increased with increasing soaking time, and it reached 20 m^2/g after firing at 1400 °C for 120 min. FE-SEM observation showed that morphology of the products obtained are composed of nanoparticles (ca. 20-50 nm).

INTRODUCTION

β-SiAlON ceramics are one of the most promising materials because it has potential for engineering applications owing to having the excellent properties, e.g. high strength and good thermal-shock resistance. β-SiAlON is a solid solution of β-Si_3N_4; a part of Si and N are replaced by Al and O. The chemical formula of β-SiAlON is indicated as follows:

$$Si_{6-z}Al_zO_zN_{8-z} \ (0 < z \leq 4.2) \qquad (1)$$

β-SiAlON ceramics are commonly produced by a reaction sintering of a mixture of Si_3N_4, AlN, and Al_2O_3 over 1500 °C. However, their corrosion resistance degrades due to glassy phase in the grein boundary [1]. Therefore, preparation of β-SiAlON powders followed by their sintering received considerable attention to avoid the formation of glassy phase [2]. In the present study, we focused on a zeolite as a raw material. Zeolites are hydrated, crystalline tectoaluminosilicates that are constructed from TO_4 tetrahedra (T = tetrahedral atom, e.g. Si and Al) [3-5]. Typical zeolites are aluminosilicate, which contain Si and Al uniformly in an atomic scale. In our previous study [6], β-SiAlON was synthesized from a mixture of zeolite and carbon by carbothermal reduction-nitridation (CRN). In recent years, gas-reduction-nitridation (GRN) process, which uses a mixture of NH_3 and hydrocarbon as reactant gases, has been proposed as one of the synthetic techniques of high-purity nitride powders [7-10]. GRN is the quite simple process since it dose not require mixing of raw materials with carbon and decarbonizing after nitriation. Moreover, formation of nitride via GRN occurres at lower temperatures than via CRN. Therefore, the objective of the present work is to synthesize high-purity β-SiAlON powder from zeolite via GRN.

EXPERIMENTAL PROCEDURE

A commercial zeolite powder (HSZ-330HUA, Tohso Chem., Tokyo) was used as a starting material. Its main characteristics are listed in Table 1. The raw zeolite has high specific surface area and is composed of fine particles about 0.6 μm. It was weighed on an alumina boat and placed in an electric furnace with a high-purity alumina tube. Then, it was fired to 700 °C at a heating rate of 5 °C/min in Ar gas (99.999% purity) in order to eliminate oxygen in the system and remove the adsorption water in zeolite. It was confirmed that the structure of zeolite did not change during the pre-heat treatment up to 700 °C. Heating was continued in a gas mixture of NH_3 (99.999% purity) and 0.5 vol% C_3H_8 (99.99% purity). The flow rate was 4 l/min. The sample was fired at 1200-1400 °C for 0-120 min before being cooled in NH_3. Furthermore, CRN process was also applied for the synthesis of SiAlON using the same raw zeolite for comparison [5].

Phases present in the products were identified by X-ray diffractometry (RINT2500, Rigaku, Tokyo, Japan) using CuKα radiation operated at 50 kV and 300 mA. Morphological changes were observed by field emission scanning electron microscopy (FE-SEM; S-5200, HITACHI, Tokyo, Japan). Specific surface areas of the powders were measured by a single-point Brunauer-Ematt-Teller method (BET; Quantasorb, Quantachrome, Boynton Beach, FL).

Table 1. Characteristics of a raw powder

Zeolite type	Y-type
Si/Al (mol/mol)	3 (z =1.5)
Mean particle size (μm)	0.6
Specific surface area (m²/g)	550

RESULTS AND DISSCUSSIONS

Figure. 1 shows XRD patterns of the products prepared by CRN and GRN methods using the same zeolite at 1400 °C for 120 min. β-SiAlON was a main phase in both samples, but XRD pattern of the β-SiAlONs phase obtained by CRN method indicated the broad and splited peak in higher diffraction angle. This fact indicates the formation of β-SiAlONs with different compositions. On the other hand, high-purity β-SiAlON with an uniform composition was confirmed in the sample produced by GRN (Fig.1(b)).

XRD patterns of the products synthesized at 1200-1400 °C for 0-60 min, together with the raw zeolite are shown in Fig. 2(a). The raw zeolite was transformed into amorphous phase at 1200 °C (Fig.2 (b)). In the sample fired at 1400 °C without soaking time, O-SiAlON and X-SiAlON were confirmed to be formed (Fig. 2(c)) similarly to a result in a previous study [10]. Finally, high-purity β-SiAlON phase was obtained by heat treatment at 1400 °C for 60 min. Preparation of β-SiAlON from zeolite by CRN requires firing at a higher temperature for longer soaking time to form β-SiAlON. Mullite phase, one of the stable phases at high temperatures was formed during nitridation process. In the case of CRN, as a result of heterogeneity of the composition, β-SiAlONs with different composition obtained. On the other hand, since GRN proceedes at lower temperature than CRN, it is possible to prevent the formation of mullite to maintain homogeneity of the chemical composition of starting materials. As a results, high-purity β-SiAlON with uniform compositon was obtained by GRN of zeolite.

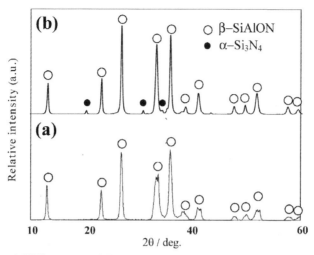

Figure 1. XRD patterns of the products (a) carbothermal reduction-nitridation (b) gas-reduction-nitridation

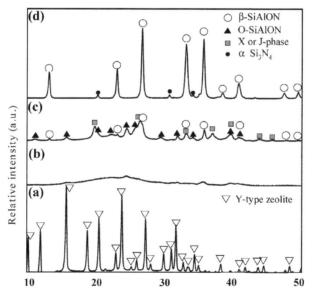

Figure 2. XRD patterns of (a) raw material and the products synthesized at (b)1200 °C for 0 min (c) 1400 °C for 0 min and (d) 1400°C for 60 min.

Fig. 3 shows FE-SEM photographs of raw zeolite and the products synthesized at 1200-1400 °C for 120 min. Morphology of the raw zeolite (Fig. 3(a)) was approximately maintained after transition to glassy phase (Fig. 3(b)). As the soaking time increases to 1400 °C, equiaxed grains with 30-50 nm (Fig. 3(c)) were obtained in diameter.

Figure 3. FE-SEM photographs of (a) the raw material and the products synthesized at (b)1200 °C for 120min (c) 1400 °C for 60 min.

Fig. 4 shows the specific surface areas of the products synthesized at various reaction temperatures for 0-60 min. Specific surface area of a raw material drastically decreased with the amorphization. However, in the sample prepared at 1400 °C, it was found that the specific surface area gradually increased up to 20 m²/g with the formation of β-SiAlON.

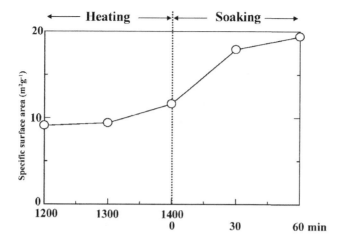

Figure 4. Specific surface area of the products synthesized at various reaction temperatures at 1400 °C for 60 min.

CONCLUSIONS

Zeolite was used as a raw material to prepare β-SiAlON by gas-reduction nitridation method. In consequence, high purity β-SiAlON powder was synthesized at 1400 °C for 60 min, and it was found that β-SiAlON obtaind are comosed of nanoparticles with diameter of about 30-50 nm. Mullite phase could not be confirmed in the samples synthesized by GRN. BET surface area of the products increased along with the formation of β- SiAlON. These promising results suggest the potentiality of gas-reduction-nitridation for synthesizing high purity β- SiAlON nanoparticles from zeolite.

REFERENCES
[1] T. Sato, T. Tokunaga, T. Endo, M. Shimada, K. Komeya, K. Nishida, M. Komatsu and T. Kameda, "Corrosion of Silicon Nitride Ceramics in Aqueous Hydrogen Chloride Solutions," *J.Mater.Sci.*, **23** 3440-3446 (1988).

[2] T.Ekström, P.O. Käll, M.Nygren and P.O.Olsson, "Dense single-phase β-SiAlON ceramics by glass-encapsulated hot isostatic pressing" *J.Mat.Sci.*, **24** 1853-61 (1989).

[3] D.W. Breck, "Zeolite Molecular Sives" *Wiley, New York* (1974).

[4] R.M. Barrer, "Hydrothermal Chemistry of Zeolites," *Academic Press, London* (1982).

[5] T. Wakihara and T. Okubo, "Hydrothermal Synthesis and Characterization of Zeolites," *Chem Llett.*, **34** 278-79 (2005).

[6] F. Li, J. Tatami, T. Meguro and K. Komeya, "Synthesis of β-SiAlON powder by Carbothermal Reduction-Nitridation of Zeolite" *Key. Eng. Mater.*, **247** 109-112 (2003).

[7] T. Suehiro, J. Tatami, T. Megro, S. Matsuo and K. Komeya, "Synthesis of Spherical AlN Particles by Gas-Reduction-Nitridation Method," *J. Eur. Ceram. Soc.*, **22** 521-26 (2002).

[8] T. Suehiro, J. Tatami, T. Megro, K. Komeya and S. Matsuo, "Aluminum Nitride Fibers Synthesized from Alumina Fibers Using Gas-Reduction-Nitridation Method," *J. Am. Ceram. Soc.*, **85** [3] 715-17 (2002).

[9] T. Suehiro, N. Hirosaki, R. Terao, J. Tatami, T. Megro and K. Komeya, "Synthesis of Aluminum Nitride Nanopowder by Gas-Reduction-Nitridation Method," *J. Am. Ceram. Soc.*, **86** [6] 1046-48 (2003).

[10] T.Yamakawa, J.Tatami, K.Komeya and T.Meguro, "Synthesis of AlN Powder from Al(OH)$_3$ by Reduction-Nitridation in a Mixture of NH$_3$-C$_3$H$_8$," *J. Eur. Ceram. Soc.*, in press, (2005).

[11] K.J.D. Mackenzie, R.H. Meinhold, G.V. White, C.M. Sheppard and B.L. Sherriff "Carbothermal Formation of β-SiAlON from Kaolinite and Halloysite by ^{29}Si and ^{27}Al Solid State MAS NMR," *J. Mat. Sci.* **29** 2611-2619 (1994).

ELECTROSPINNING OF CERAMIC NANOFIBERS AND NANOFIBER COMPOSITES

Junhan Yuh, Hyun Park and Wolfgang M. Sigmund*

Department of Materials Science & Engineering, University of Florida,

225 Rhines Hall, P.O. Box 116400, Gainesville, FL 32611, USA

ABSTRACT

Lanthanum cuprate nanofibers and barium titanate nanofibers were synthesized by electrospinning utilizing sol-gel precursors. The effects of electrospinning parameters (ceramic precursor concentration and electric voltage) on fiber diameter before/after calcinations were investigated in detail for lanthanum cuprate nanofibers. A series of experiments designed by response surface methodology shows as dried fiber diameter increases calcined fiber diameter increases. As voltage decreases and ceramic precursor concentration increases, dried fiber diameter increases for most combinations of voltage and precursor concentration. But there are also conditions available where the opposite is true. This paper explains how contradicting reports on fiber diameter dependence in the literature can be unified in a larger picture. Furthermore, the microstructure evolution in the nanofiber was studied. The morphology, microstructure and crystal structure were investigated by SEM, TEM and XRD.

INTRODUCTION

Over the last few decades, research and development of one dimensional nano materials such as nanotubes and nanofibers have been of great interest due to their unique structure and properties, i.e. high aspect ratio, large specific surface area and chemical / mechanical stabilities that can be utilized for diverse applications in the field of nano devices. Nanomaterials can be used as building blocks in nanotechnology.[[1, 2]] Currently, several ceramic nanowires have been developed by various techniques, e.g. template growth, laser ablation and chemical vapor deposition (CVD)[2-5]. But a major disadvantage of these methods is the need of multiple processing steps to produce nanowires. Etching for the template removal and purification processes are good examples of such extra processing steps required as they are both time and cost consuming processes. [1, 6]

A few years ago, research on electrospinning of ceramic fibers started since this process provides direct access to nanostructured ceramic nanofibers and requires less steps compared to other methods. In electrospinning of ceramic nanofibers, a high electric field is applied to a metallic syringe needle containing sol-gel precursors mixed with dissolved polymer. Under the influence of the electric field, a droplet of the polymer solution at the needle tip is deformed into a conical shape. If the electrostatic repulsive force overcomes the surface tension, a thin charged jet is ejected. The jet moves towards a grounded plate. Due to the presence of polymer entanglements, the jet remains stable and does not transform into spherical droplets. As the solvent evaporates, this jet is stretched to many times its original length to produce continuous, ultrathin fibers of polymer. Recently, binary ceramic nanofibers such as titania, vanadium oxide, niobium oxide, silica and copper oxide have been produced via electrospinning, [7-14] and a few ternary ceramic nanofibers were fabricated.[15, 16] Electrospun nanofibers come naturally with a high surface to volume ratio and are interesting for applications in catalysis and as electrodes in solid oxide fuel cells (SOFCs) with lanthanum cuprate being one of the parent materials for its

*To whom correspondence should be addressed: Tel: 1-352-846-3338, Fax: 1-352-846-3355, e-mail: wsigm@mse.ufl.edu

9

cathode.[17] SOFCS offer the advantage of high energy conversion efficiency combined with low pollution for power generation. Lanthanum cuprate that is doped with divalent ions such as strontium, forms an oxygen deficient perovskite structure.[18] This yields a perovskite oxide that is electrically conductive, which made it a promising candidate for the electrode material for SOFCs [19] and gas sensors.[20] We will compare the results found for electrospun lanthanum cuprates with another electrospun perovskite, i.e. barium titanate. This ferroelectric material is widely used in various electronic applications and with the current miniaturization trend; development of nanoscale materials with ferroelectricity is crucial.

Since electrospinning is a novel technique for ceramics only few reports have been published to now. Therefore, we present a systematic study of the ceramic precursor concentration on the fiber diameter. We also report on the electrospinning synthesis of two ternary perovskite oxide nanofibers, lanthanum cuprate and barium titanate. Results will be presented in empirical models that describe effects of variables (ceramic precursor concentration, electric voltage) on fiber diameters. Parameter optimization and derivation of empirical models was based on response surface methodology.[21, 22]

EXPERIMENTAL

The chemicals used for the electrospinning solution were lanthanum nitrate hydrate and copper nitrate, and polyacrylamide (M_w=10,000, 50 wt.% solution in water) for lanthanum cuprate. Barium titanate precursors were chosen as barium acetate, titanium isopropoxide and poly(vinyl pyrrolidone) (PVP, M_w=1,300,000). All chemicals were used as received. Electrospinning solutions were prepared by mixing lanthanum nitrate hydrate, copper nitrate and polyacrylamide with stirring for 24 hours. Ceramic precursor concentrations of lanthanum cuprate were determined by experimental design (Table 1). The solutions were taken into 5 ml syringes and pumped out of the syringe using a syringe pump (BSP 99, Braintree Scientific Inc., Braintree, USA). The distance between the metal needle and collector (grounded aluminum foil) was 5cm. A voltage according to the experimental design (Table 1) was applied to the metal needle and fibers were collected on an aluminum foil. Fibers were dried for 12 hours at 100°C, followed by calcination for 10 hours at 600°C.

For the preparation of barium titanate via sol-gel processing, barium acetate and titanium isopropoxide were dissolved in acetic acid at a 1:1 molar ratio. After dissolution of barium acetate and titanium isopropoxide, PVP as solution in ethanol (PVP 0.2g and ethanol 3mL) was added and everything stirred at room temperature for 2 h. This solution was loaded into the syringe. Electrospinning was conducted by applying an electric field of 2,143 V to the precursor mixture. As-synthesized non-woven mats of nanofibers were dried in a vacuum oven at 80°C for an hour. For crystallization an annealing process with holding temperatures of 650°C and 750°C for 60 minutes was applied. Morphologies and dimensions of fibers were investigated using field emission scanning electron microscopy (JSM-6335F, JEOL, Peabody, USA) with an acceleration voltage of 15kV. X-ray diffraction (APD 3720, Philips, New York, USA) was used to characterize the phases. Transmission electron microscopy operating at 200kV was utilized for microstructural analysis.

Solutions were characterized for their surface tension by the Wilhelmy plate method and solution viscosities were measured with a rheometer (MCR 300, Paar Physica USA, Ashland, USA). Measurements were done at room temperature. The solution was pre-sheared at 800s^{-1} for one minute, then, kept stationary for one minute to equilibrate and then, the shear rate was ramped from $0.001s^{-1}$ to $1100s^{-1}$. The time duration between two measurements was 15 seconds. Electrical conductivities were measured with a conductivity meter (AB30, Fisher Sci. Fairlawn, USA). A series of experiments was designed using two variables (ceramic precursor concentration, electric voltage). Experimental conditions were determined according to D-optimal design. This design selected design points that minimize the variance of the model regression coefficients. To derive quadratic equations, at least 3 levels of experimental conditions for each factor were investigated. Designed experimental conditions are given in Table 1.

Table 1. Experimental conditions.

Experiment #	$Conc.$ (M)	$U_{applied}$ (kV)
1	0.1	25
2	0.01	10
3	0.055	17.5
4	0.01	25
5	0.1	10
6	0.1	17.5
7	0.1	25
8	0.01	17.5
9	0.055	25
10	0.01	10
11	0.1	10
12	0.055	10

RESULTS

1. Lanthanum cuprate

1.1 Response surface methodology for dried fiber diameter

Fig. 1 shows scanning electron microscopy (SEM) images of dried fibers and heat-treated fibers. They are typical for electrospun ceramic nanofibers. As synthesized dried fibers are nanocomposites of the sol-gel ceramic and the polymer. On heating, the polymer burns out and the microstructure of the nanofibers develops. The fiber ceramic microstructure can be controlled by the annealing process, but not the fiber diameter. For fiber diameter quantification at least 50 fiber diameters were measured on several SEM pictures by randomly picking fibers and measuring their diameter vertically to their length.

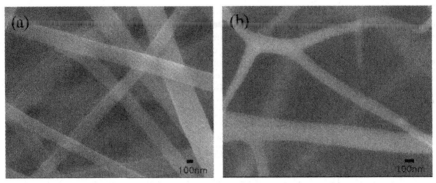

Fig 1: SEM pictures of (a) dry as-synthesized lanthanum cuprate/polyacrylic acid composite nanofibers, (b) lanthanum cuprate nanofibers after annealing (600°C for 1h)

Measurement results for each experimental condition are summarized in Table 2. Validities F of the model and regression coefficients are summarized in Table 3 and Table 4, respectively.

$$D_{dried\ fiber}\ (nm) = 188.4 - 32.2 \times V + 78.3 \times C - 22.5 \times V^2 + 52.9 \times C^2 - 35.1 \times V \times C \qquad (1)$$

where $D_{dried\ fiber}$ is the dried fiber diameter in nm, V is the coded variable for voltage with

$$V = (U_{applied} - 17.5) / 7.5 \qquad (2)$$

where $U_{applied}$ is the applied potential in kV. C is the coded variable for the ceramic precursor concentration defined as C=Conc.-0.055/0.045 with Conc. using molar units.

It follows from equation (1) that the ceramic precursor concentration is more important to determine the dried fiber diameter than $U_{applied}$. Furthermore it clearly states that as $U_{applied}$ decreases and ceramic precursor concentration increases, fiber diameter increases. Additionally, there is an interaction between voltage and ceramic precursor concentration, which is more effective than voltage alone. This can be seen in Fig. 2 where a contour plot of the dried fiber diameter as a function of voltage and ceramic precursor concentration is presented. Two main areas can be seen. In one area where 0.01 M < *Conc.* < 0.03 M $D_{dried\ fiber}$ increases with increasing $U_{applied}$ up to 21 kV .[11]

Table 2: Coded Variables C and V for conc. and voltage, respectively, with the resulting fiber diameters for dried and calcined lanthanum cuprate nanofibers.

Experiment #	C	V	$D_{dried\,fiber}$ (nm)	$D_{calcined}$ (nm)
1	1	1	226	62
2	-1	-1	138	86
3	0	0	186	52
4	-1	1	145	84
5	1	-1	351	156
6	1	0	319	103
7	1	1	233	54
8	-1	0	162	70
9	0	1	135	78
10	-1	-1	141	96
11	1	-1	378	159
12	0	-1	206	89

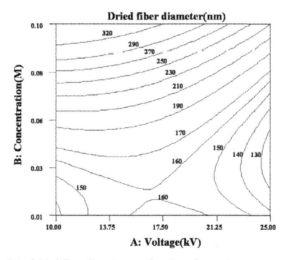

Fig. 2: Contour plot of dried fiber diameter as a function of ceramic precursor concentration and voltage for lanthanum cuprate nanofibers.

1.2 Response surface methodology for calcined fiber diameter

For calcined fibers the same methodology was applied as for dried fibers. Measurement results of each condition are also summarized in Table 2. A SEM image of calcined lanthanum cuprate fibers is shown in Fig. 1(b). An analysis was done to confirm the model. The regression coefficients and the validity F for each model are summarized in Table 3 and Table 4, respectively.

Table 3: Validity F of the models for lanthanum cuprate nanofibers

Model	F	R^2	Adj. R^2
Dried fiber model	73.21	0.948	0.934
Calcined fiber model	12.63	0.76	0.699

Table 4: Regression coefficients of each model for lanthanum cuprate nanofibers

Variable	Dried fiber model	Calcined fiber model
$U_{applied}$	-32.29	-19.32
Conc.	78.3	16.63
V^2	-22.52	9.29
C^2	52.94	16.3
CxV	-35.13	-20.91

For calcined fibers the following equation (3) is found:

$$D_{calcined}\ (nm) = 70.95 - 19.32 \times V + 16.63 \times C + 9.29 \times V^2 + 16.30 \times C^2 - 20.91 \times V \times C \qquad (3)$$

where $D_{calcined}$ is the calcined fiber diameter in nm. C is the coded variable for the ceramic precursor concentration defined as $C=Conc.-0.055/0.045$ with $Conc.$ using molar units, and V is the coded variable for voltage as mentioned earlier.

Equation (3) has similarities to equation (1) and some differences. It is similar in that the ceramic precursor concentration is more important than voltage to determine the calcined fiber diameter, and as voltage decreases and ceramic precursor concentration increases, fiber diameter increases. There is also an interaction between voltage and ceramic precursor concentration, which is more important. The difference, however, is found in that there is only a very small region where calcined fiber diameter increases with increasing voltage (25 kV and 0.01 M). The major difference is a new saddle point. It is displayed in the contour plot of the calcined fiber diameter as a function of voltage and ceramic precursor concentration in Fig. 3. This saddle point can be used to minimize the calcined fiber diameter.

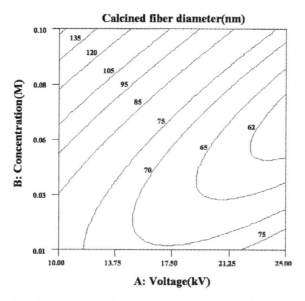

Fig. 3: Contour plot of calcined fiber diameter as a function of ceramic precursor concentration and voltage for lanthanum cuprate nanofibers

2. Barium titanate

The second electrospun ceramic nanofiber material that is presented is a ferroelectric. Here, we focus on the microstructure evolution in the nanofiber with heat treatment. The X-ray diffraction (XRD) patterns of electrospun barium titanate nanofibers with heat treatment at different temperature are shown in Fig. 4(a). The XRD pattern of as-synthesized and dried at 80°C sample shows their amorphous nature. After heat treatment at 650°C the onset of perovskite $BaTiO_3$ is evident, although there are some non-perovskite peaks (~27°, 2θ). Non-perovskite peaks were due to the complex structural chemistry of barium titanate and at this temperature, 650°C, perovskite and non perovskite peaks are coexist. XRD pattern obtained from fibers heat treated at 750°C shows well-defined, high intensity perovskite barium titanate peaks with no detectable non-perovskite phases. The increment of the intensity of the peaks corresponding to temperature increment indicates that amount of the tetragonal barium titanate phase was increased.

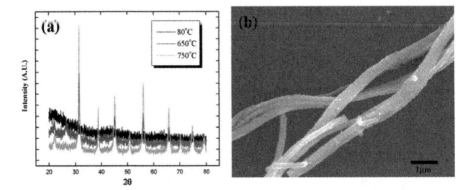

Fig. 4: (a) XRD patterns of barium titante nanofibers heat treated at various conditions: 80°C, 650°C and 750°C (b) SEM image of barium titanate nanofibers after heat treatment at 750°C for 1 h.

For comparison reasons to the study on fiber diameter of lanthanum cuprates, SEM images of electrospun barium titanate after heat treatment at 750°C were done and quantitatively analyzed (see Fig. 4(b)). Fiber diameters were measured for 50 fibers of both dry as-synthesized and annealed (750°C) nanofibers. Average fiber diameter before heat treatment was found to be 273 nm and after heat treatment at 750°C the average diameter was reduced to 116 nm. This is equivalent to 60% of diameter reduction upon annealing. The reduction can mainly be attributed to polymer burn out during the heat treatment. The shrinkage in fiber diameter does not directly correspond to the above presented equations (1) and (3) since a different polymer has been used with higher molecular weight thus influencing surface tension and viscosity. Furthermore, the different precursor concentrations as well as the ionic constituents for barium titanate also alter the bulk nanofiber conductivity. This demonstrates that the models for one ceramic system cannot easily be transferred to another ceramic material.

During annealing the material crystallizes and for $BaTiO_3$ equiaxed grains of 25-50 nm diameter become visible. They were verified to be perovskite barium titanate by transmission electron microscopy (TEM) (see figure 5) and selected area diffraction, which is published elsewhere.[29] The upper right inset is a high resolution image of an individual grain, which shows the lattice fringes. This confirms the crystallinity of the individual grains.

Fig. 5: TEM image of barium titanate nanofiber annealed at 750°C for 1 h. Right upper inset: high resolution image of barium titanate nanofiber showing lattice fringes with scale bar of 2 nm.

DISCUSSION

1. Process optimization

For applications in nanotechnology the knowledge for nanofiber minimization is crucial. For the investigated system based on lanthanum cuprate a minimum for calcined fibers was found. When the highest voltage (25kV) is paired with a medium ceramic precursor concentration (0.07M) the thinnest calcined nanofibers can be fabricated yielding dried fibers with D_{dried} = 152 nm, and calcined fibers with diameters of 61 nm.

2. Effects of voltage

Up to this paper, the influence of electrospinning voltage on fiber diameter had not been clearly established.[23] A previous study showed that fiber diameter decreased with increasing voltage [24, 25] while others showed the opposite trend[26]. The contour map presented in Fig. 2 shows and explains both trends simultaneously. It is obvious that both observations are true. At a low voltage, the fiber diameter increases with increasing voltage. The relationship reverses at 17 kV, now the fiber diameter decreases with increasing voltage.

3. Effects of ceramic precursor concentration

As seen in the presented contour a plot, there is an impact of the ceramic precursor concentration on the final nanofiber diameter. These differences can be explained by a recently introduced model.[27] Table 5 lists the main parameters that control fiber diameter from a fundamental physics point of view. It is found that as the ceramic precursor concentration increases conductivity increases due to the increase in ionic strength. Most groups reported that increasing conductivity decreases the fiber diameter. They reported that salts such as sodium chloride and lithium chloride were used to increase the conductivity. These salts ionize when they are added to electrospinning solution and increase the conductivity through the bulk of the fiber. In our systems the conductivity of the ceramic precursor solutions is typically 3 orders of magnitude higher than for pure polymer solutions. [11, 23, 28] Thus, fiber diameter increases with increasing ceramic precursor concentration for most conditions.

Table 5: Viscosity, conductivity and surface tension data as a function of ceramic precursor concentration for lanthanum cuprate.

Conc. (M)	Viscosity (Pa s)	Conductivity (mS/cm)	Surface tension (mN/m)
0.01	0.14	11.69	76.24
0.055	0.13	12.09	77.2
0.1	0.12	18.96	78.96

SUMMARY

Lanthanum cuprate nanofibers and barium titanate nanofibers were synthesized by electrospinning. A series of experiments designed by response surface methodology shows that the ceramic precursor concentration is a more important factor than voltage to determine the response (fiber diameter). As voltage decreases and ceramic precursor concentration increases, the dried fiber diameter is found to increase. Calcined fiber diameters show the same trend.

Using the derived equations, the processes were optimized to produce dried fiber diameters of 152 nm, and calcined fiber with diameters of 61 nm at the highest voltage (25 kV) and medium ceramic precursor concentration (0.07 M).

BaTiO$_3$ fibers were found to form perovskite 150°C above the standard reported temperature for sol-gel derived BaTiO$_3$. Electrospun barium titanate was still amorphous at 600°C which might be caused by the strong electric field. Since the crystallographic study of nanofibers confirmed tetragonal perovskite barium titanate after heat treatment at 750°C, ferroelectric behavior can be expected. To confirm ferroelectricity of electrospun barium titanate nanofibers, scanning probe microscopy (SPM) and electrostatic force microscopy (EFM) investigations are ongoing.

ACKNOWLEDGEMENTS

The authors acknowledge funding from the Department of Energy under grant no. DE-FG26-03NT41614.

REFERENCES

1. Y. Dzenis, "Spinning continuous fibers for nanotechnology," *Science,* **304**(5679), 1917-1919 (2004).
2. C.N.R. Rao, A. Govindaraj, G. Gundiah, and S.R.C. Vivekchand, "Nanotubes and nanowires," *Chem. Eng. Science,* **59**(22-23) 4665-4671 (2004).
3. S. Iijima, "Helical Microtubules of Graphitic Carbon," *Nature,* **354**(6348), 56-58 (1991).
4. A.M. Morales, and C.M. Lieber, "A laser ablation method for the synthesis of crystalline semiconductor nanowires," *Science,* **279**(5348), 208-211 (1998).
5. C.R. Martin, "Nanomaterials - a Membrane-Based Synthetic Approach," *Science,* **266**(5193), 1961-1966 (1994).
6. D. Li, Y.L. Wang, and Y.N. Xia, "Electrospinning of polymeric and ceramic nanofibers as uniaxially aligned arrays," *Nano Lett.,* **3**(8), 1167-1171 (2003).
7. K.H. Lee,H.Y. Kim, M.S. Khil, Y.M. Ra, and D.R. Lee, "Characterization of nano-structured poly(epsilon-caprolactone) nonwoven mats via electrospinning," *Polymer,* **44**(4), 1287-1294 (2003).
8. S.A. Theron, A.L. Yarin, E. Zussman, and E. Kroll, "Multiple jets in electrospinning: experiment and modeling," *Polymer,* **46**(9), 2889-2899 (2005).
9. A.L. Yarin, and E. Zussman, "Upward needleless electrospinning of multiple nanofibers," *Polymer,* **45**(9), 2977-2980 (2004).
10. A. Frenot, and I.S. Chronakis, "Polymer nanofibers assembled by electrospinning," *Curr. Opin. Coll. Inter. Sci.,* **8**(1), 64-75 (2003).
11. Z.M. Huang, Y.Z. Zhang, M. Kotaki, and S. Ramakrishna, "A review on polymer nanofibers by electrospinning and their applications in nanocomposites," *Comp. Sci. Tech.,* **63**(15), 2223-2253 (2003).
12. J.M. Deitzel, J. Kleinmeyer, D. Harris, and N.C.B. Tan, "The effect of processing variables on the morphology of electrospun nanofibers and textiles," *Polymer,* **42**(1) 261-272 (2001).
13. M.M. Demir, I. Yilgor, E. Yilgor and B. Erman, "Electrospinning of polyurethane fibers," *Polymer,* **43**(11), 3303-3309 (2002).
14. X.H. Zong, K. Kim, D.F. Fang, S.F. Ran, B.S. Hsiao, and B. Chu, "Structure and process relationship of electrospun bioabsorbable nanofiber membranes," *Polymer,* **43**(16), 4403-4412 (2002).
15. H.Y. Guan, C.L Shao, Y.C. Liu, N. Yu, and X.H. Yang, "Fabrication of NiCO2O4 nanofibers by electrospinning," *Solid State Commun.,* **131**(2), 107-109 (2004).
16. D. Li, T. Herricks, and Y.N. Xia, "Magnetic nanofibers of nickel ferrite prepared by electrospinning," *Appl. Phys. Lett.,* **83**(22), 4586-4588 (2003).
17. N. Dharmaraj, H.C. Park, D.M. Li, P. Viswanathamurthi, H.Y. Kim, and D.R. Lee, "Preparation and morphology of magnesium titanate nanofibres via electrospinning," *Inorg. Chem. Comm.,* **7**(3), 431-433 (2004).
18. H.C. Yu, and K.Z. Fung, "$La_{1-x}SrCuO_{2.5}$-delta as new cathode materials for intermediate temperature solid oxide fuel cells," *Mat. Res. Bull.,* **38**(2), 231-239 (2003).
19. H.C. Yu, and K.Z. Fung, "Electrode properties of $La_{1-x}SrxCuO_{2.5}$-delta as new cathode materials for intermediate-temperature SOFCs.," *J. Power Sources,* **133**(2), 162-168 (2004).

20. X.H. Zhou, Q.X. Cao, Y. Hu, J.X. Gao, and Y.L. Xu, "Sensing behavior and mechanism of La₂CuO₄-3uO₂ gas sensors," Sen. Actuators B., 77(1-2), 443-446 (2001).

21. D.C. Montgomery, "Design and analysis of experiments," 5th ed., New York: *John Wiley*. xii, 684 (2001).

22. R.H. Myers, and D.C. Montgomery, "Response surface methodology : process and product in optimization using designed experiments," Wiley series in probability and statistics. Applied probability and statistics, New York: *Wiley*. xiv, 700 (1995).

23. D. Li, and Y.N. Xia, "Electrospinning of nanofibers: Reinventing the wheel?," *Adv. Materials*, **16**(14), 1151-1170 (2004).

24. Y.M. Shin, M.M. Hohman, M.P. Brenner and G.C. Rutledge, "Experimental characterization of electrospinning: the electrically forced jet and instabilities," *Polymer*, **42**(25), 9955-9967 (2001).

25. J.S. Lee, K.H. Choi, H.D. Ghim, S.S. Kim, D.H. Chun, H.Y. Kim and W.S. Lyoo, "Role of molecular weight of atactic poly(vinyl alcohol) (PVA) in the structure and properties of PVA nanofabric prepared by electrospinning," J. *Appl. Poly. Sci.*, **93**(4), 1638-1646 (2004).

26. S.L. Zhao, X.H. Wu, L.G. Wang, and Y. Huang, "Electrospinning of ethyl-cyanoethyl cellulose/tetrahydrofuran solutions," J. *Appl. Poly. Sci.*, 2004. **91**(1): p. 242-246.

27. H. Park, "Fabrication of lanthanum copper oxide nanofibers by electrospinning," Ph. D dissertation, *University of Florida* (2005)

28. W.K. Son, J.H. Youk, T.S. Lee, and W.H. Park, "The effects of solution properties and polyelectrolyte on electrospinning of ultrafine poly(ethylene oxide) fibers," *Polymer*, **45**(9), 2959-2966 (2004).

29. J.H. Yuh, J. Nino, W. Sigmund, "Synthesis of barium titanate (BaTiO3) nanofibers via electrospinning, Materials Letters, in press, **2005**

MELT SYNTHESIS AND CHARACTERIZATION OF $(A_{1-x}A'_x)(B_{1-y}B'_y)O_3$ COMPLEXED OXIDE PEROVSKITES

Tadashi Ishigaki

Kazumasa Seki

Shunji Araki

Naonori Sakamoto

Tomoaki Watanabe

Masahiro Yoshimura

Materials and Structures Laboratory, Tokyo Institute of Technology

4259 Nagatsuta, Midori, Yokohama 226-8503, Japan

E-mail: yoshimura@msl.titech.ac.jp

ABSTRACT

In order to synthesize various Perovskite ABO_3 type compounds and their solid solutions, we have applied a novel "melt synthesis technique" rather than conventional solid state reaction techniques. In the melt synthesis, the mixture of oxides or their precursors is melted in a short period of time (1-60 sec) by a strong light radiation in an arc imaging furnace. A spherical molten sample where multiple cations were mixed homogeneously was directly solidified on a copper hearth with a rapid cooling of 10^2K/sec. $LaAlO_3$, $GdScO_3$, $ATiO_3(A = Ba, Sr, Ca$ and $La)$ and the solid solutions among them, $(A_{1-x}A'_x)(B_{1-y}B'_y)O_3$, were synthesized in this technique. Madelung potentials was calculated by MADEL, and discussed about lattice energy of these components.

INTRODUCTION

Perovskite ABO_3 type compounds have been investigated in wide area of fields like ferroelectrics, magnetics, optics, phosphors, electronics, ionics, superconductors, sensors, catalysts, etc., because they have various but unique properties due to their compositions and structures which are characterized by the various combinations of A and B carions. In order to form the perovskite and/or related structures, A and B ions should have following conditions: [1][2]

1. A-ion is sufficiently larger than B-ion in size.
2. The sum of the valences of A-ion and B-ion must be six.
3. The valence of B-ion is larger than that of A-ion.

Under these conditions, a lot of compounds with the formula of $A^{3+}B^{3+}O_3$, $A^{2+}B^{4+}O_3$, $A^+B^{5+}O_3$, even $B^{6+}O_3$ and their solid solutions can be included; that is, "Complex Perovskites" where multiple A cations and B cations are involved in the lattice. The synthesis of such complex perovskites is not easy by conventional solid state reaction techniques because the reaction rates among oxides are so slow by solid state diffusion as to form homogeneous compounds or solid solutions.

Melt synthesis is rapid and homogeneous, because ions are mixed homogeneously in the melt. However, anions and cations may exist randomly in the melt. Generally, coordination number of A site cation is 12, and B site is 6. Even in the melt, the local combination of anions and cations seems to be same. When anions and cations exist completely random, big volume changed was expected when the melt was crystallized, but such a large volume change was not observed. It means that, each ions isolate in "plasma model", but not in "melt model".

The melt synthesis is a high temperature approach for the synthesis which is rather opposite from solid state syntheses where a low temperature approach to the synthetic temperature. However, the melt synthesis has not widely been studied yet probably because it is not easy to make such a high temperature and to get a non-reactive container to the molten oxides at such a high temperature of 1500-2000°C. We have succeeded to synthesize series of perovskite-type phases in the systems of $GdScO_3$-$BaTiO_3$, -$SrTiO_3$, -$CaTiO_3$, $LaAlO_3$-$BaTiO_3$, -$SrTiO_3$, -$CaTiO_3$, etc., which are the solid solutions in the systems of $A^{2+}B^{4+}O_3$-$A^{3+}B^{3+}O_3$. Their lattice parameters and lattice distortions could be evaluated in the values of sizes and valences in A and B ions. Structural transformations in ABO_3 have also been studied as function of temperatures and in relation to kinds of A and B ions. Some of them may have metastable states because of quenching from high temperatures in the melt synthesis. Therefore, we have expected to synthesize metastable perovskite phases.

Combining advantages of Melt Synthesis and Arc Imaging Furnace, we studied to prepare Perovskite solid solutions rapidly.

EXPERIMENT

Arc imaging furnace

This furnace is designed to heat small-size samples upto very high temperature like above 2000°C to melt in clean conditions. As shown in the Figure 1 the light emitted from a 10 Kw Xenon Lamp are collected by an ellipsoidal mirror (Collector), reflected by a plane mirror and then emitted by another ellipsoidal mirror (Emitter) to the sample. A molded sample with 3-8 mm size of mixed powders was put on a water-cooled Cu hearth. The sample can be heated in various atmospheres in a very short period of time like a few seconds and also cooled down by breaking the light radiation by a shutter or quick removing the sample stage from the focus of the emitter mirror. The brightness temperature can be monitored by a pyrometer. To calibrate this pyrometry system, the solidification temperatures of refractory oxides have been measured as Al_2O_3 (2054±4°C),

Y_2O_3 (2433±3°C), HfO_2 (2803±3°C), and CaO (2899±3°C). [3, 4] Details of the furnace and the pyrometry have been reported previous papers [2, 3, 4]. When a sample is melted, it becomes a small molten globule generally with 1-5mm in size by its surface tension, thus it gives almost point-contract to the Cu hearth, which make minimize the heat loss by the hearth. When the bottom of the sample where unmolten, the sample can be remelted after its turn over.

Sample preparation

Starting materials were high purity oxides (TiO_2 (99.99%, Kojundo Chemical Laboratory Co., Ltd.), $SrCO_3$ (99.9%, Kojundo Chemical Laboratory Co., Ltd.), $BaCO_3$ (99.95%, Kojundo Chemical Laboratory Co., Ltd.), $CaCO_3$ (99.5%, Kanto Kagaku), Al_2O_3 (99.99%, AKP-30, Sumitomo Chemical Co., Ltd), Gd_2O_3 (99.99%, Shin-Etsu Chemical Co., Ltd.), La_2O_3 (99.99%, Shin-Etsu Chemical Co., Ltd.), and Sc_2O_3 (Mitsubishi Metal Coporation)) powders. They were mixed by dry and wet mixing in a high-purity alumina mortar. The mixed powder or their molded samples were placed on a copper hearth and melted in vacuum using the arc-imaging furnace. It took generally 5-15 seconds from a molten state at around 2000°C to a solid state with dark color at around 600°C. The cooling rate was, therefore, estimated to be more than 100K/sec.

All the samples were characterized by X-ray diffraction. Powder X-ray diffraction patterns were obtained from arc-melted samples after grinding using an alumina mortar, with CuKα radiation in a curved graphite-beam monochrometer (MXP3VA, MAC Science, Tokyo, Japan).

RESULTS AND DISCUSSION

$LaAlO_3$, $GdScO_3$, $CaTiO_3$, $SrTiO_3$, and $BaTiO_3$ could be synthesized by the melt method in short periods of time like 5-60 seconds in a single step from the mixed powders. The crystal structural of $GdScO_3$ and $CaTiO_3$ was orthorhombic-Perovskites. And $LaAlO_3$, $BaTiO_3$ $SrTiO_3$ was a perovskite of rhombohedral, hexagonal, and cubic respectively. As shown in Figure 2, a pure $LaAlO_3$ sample could be synthesized by this method in a short period of total 15 sec, 5 sec×3 times. It is noticeable merit of the melt synthesis because solid state synthesis would generally take multi steps: molding to a pellet, heating at high temperatures and grinding, repeatedly. For example, pure sample of $LaAlO_3$ [5] has been synthesized by heating at 1700°C for 2h followed several intermediate grindings and moldings the calcined again at 1500°C.

In the case of Ba-Ti-La system oxide, it takes more than 10 hours to prepare. [6, 7]

The melt synthesis might prevent the contamination from grinding media and containers during grindings. We have also succeeded to synthesize in a single step multiple series of Perovskite type solid solutions for the systems among $LaAlO_3$, $GdScO_3$, $BaTiO_3$, $SrTiO_3$, and $CaTiO_3$.

Except for the $GdScO_3$-$LaAlO_3$ systems, all the samples were the single phases. Figure 3 shows unit cell volumes per cation of Perovskite solid solution phases. Continuous volume changes suggest continuous solid solutions in spite of different crystal systems in Perovskites: that is, $A^{2+}B^{4+}O_3$ and $A^{3+}B^{3+}O_3$. On the other hand, in the system of $GdScO_3$-$LaAlO_3$, continuous solid solutions were not obtained even though $GdScO_3$ and $LaAlO_3$ are belonged in the isovalent Perovskites of $A^{3+}B^{3+}O_3$. The products were separated into two phases, $LaAlO_3$-rich trigonal and $GdScO_3$- rich orthorhombic phases. The reason is not clear in this moment, but one reason seems to be a large difference between the ionic radios of Sc^{3+} and Al^{3+}, i.e. ionic radios of Sc^{3+} and Al^{3+} is 0.745 and 0.535, respectively. [8] Each lattice energies were calculated by assuming appropriate the crystal structural parameters from ICSD database and calculation done by the MADEL [9] program. $LaAlO_3$ [10] and $GdScO_3$ [11] are -3903 kcal/mol and 4199 kcal/mol. $BaTiO_3$(hexagonal) [12], $SrTiO_3$ [13] and $CaTiO_3$ [14] are -4106, -4219, -4236 kcal/mol.

We have tried to expand the melt synthesis technique to study other system $BaTiO_3$-$1/2La_2Ti_2O_7$. ABO_3-Perovskite and $A_2B_2O_7$-Pyrochlore have similar in cationic composition, where the atomic ratio of A and B is 1:1. Figure 4 shows X-ray diffraction patterns of the products in the $BaTiO_3$-$1/2La_2Ti_2O_7$ system. In this system, continuous solid solutions have not been formed, but four phases were found, that is, a Pyrochlore phase in $1/2La_2Ti_2O_7$, $0.1BaTiO_3$-$0.45La_2Ti_2O_7$, and $0.2BaTiO_3$-$0.4La_2Ti_2O_7$ samples, a tetragonal-Perovskite phases in $0.6BaTiO_3$-$0.2La_2Ti_2O_7$ and a hexagonal-$BaTiO_3$ samples, and an intermediate compound of $BaLa_2Ti_3O_{10}$ in the sample ($0.4BaTiO_3$-$0.3La_2Ti_2O_7$). Those Pyrochlore were (h 0 0)-orientated, $BaLa_2Ti_3O_{10}$ was (0 k 0)-orientated. Figure 6 shows SEM image of melted sample $BaLa_2Ti_3O_{10}$ crashed. It is observed cleavage surface. The compound of $BaLa_2Ti_3O_{10}$ has been reported by Skapin et al. [15] in the system BaO-La_2O_3-TiO_2 (Figure 5). In order to explain the absent of $LaTiO_3$ and present of $La_2Ti_2O_7$, lattice energies of these compounds were calculated using MADEL as same as described above. According to the results, $LaTiO_3$ [16] lattice energy U_{LaTiO3} is -4391 kcal/mol (-2196 kcal/cation), $La_2Ti_2O_7$ [17] lattice energy $U_{La2Ti2O7}$ is -9928 kcal/mol (-2482 kcal/cation). It implies that $La_2Ti_2O_7$ more easy to form than $LaTiO_3$. Similarly, lattice energies for $BaTiO_3$ (tetragonal)[18] and $BaLa_2Ti_3O_{10}$ [19] were calculated. The values are U_{BaTiO3} = -4111 kcal/mol (-2056 kcal/cation), $U_{BaLa2Ti3O10}$ = -15398 kcal/mol (-2566 kcal/cation) respectively. Obtained values suggested that the most stable phase is $BaLa_2Ti_3O_{10}$, because of $U_{BaLa2Ti3O10}$ < U_{BaTiO3} + $U_{La2Ti2O7}$.

In fact, we can observe the diffraction peaks corresponding to $BaTiO_3$ in the XRD pattern for the sample with this system. Those results seem to indicate that the melt synthesis is useful to study for the systems including multiple phases. The formation of continuous solid solution requires the similarity in crystal structures between two end members.

CONCLUSION

From a scientific and engineering point of view, Melt-solidification method is suitable for material researches in high temperature syntheses.

Various Perovskite compounds and their solid solutions in the systems of $(A_{1-x}A'_x)(B_{1-y}B'_y)O_3$ Perovskite were prepared by the rapid synthesis via melt-solidification methods using the arc imaging furnace. On the other hands, the single phase of solid solutions were not formed in the system of $ABO_3 - 1/2A'_2B'_2O_7$. Because of the stable lattice energy, the intermediate compound $BaLa_2Ti_3O_{10}$ was formed in the system $BaTiO_3 - 1/2La_2Ti_2O_7$.

REFERENCES

[1] F. S. Galasso, Structure, properties, and preparation of perovskite-type compounds, (Oxford ; New York : Pergamon Press), (1969)

[2] M. Yoshimura, Bull. Tokyo Inst. Tech., 120 (1974) 13-27

[3] Toyoaki Yamada, High temp. High press., 18 (1986) 377-388

[4] Toyoaki Yamada, Masahiro Yoshimura, Shigeyuki Somiya, J. Am. Ceram. Soc., 69 (1986) C-243-245

[5] M. Kakihana and T. Okubo J. Alloys and Compounds, 266 (1998) 129-133

[6] J. Takahashi, T. Ikegami and K. Kageyama, J. Am. Ceram. Soc., 74 (1991) 1873-1879

[7] J. P. Guha, J. Am. Ceram. Soc., 74 (1991) 878-880

[8] R. D. Shannon, Acta Cryst., A32 (1976) 751-767

[9] K. Kato, FORTRAN77 program to calculate electrostatic site potentials and Madelung enegies of ionic crystals by Fourier method, The Rietveld Programs "RIETAN " F. Izumi (1991)

[10] ICSD#74494

[11] ICSD#99543

[12] ICSD#75240

[13] ICSD#56717

[14] ICSD#16688

[15] S. Skapin, D. Kolar, D. Suvorov, Z. Samardzija, J. Mater. Res., 13 (1998) 1327-1334

[16] ICSD#98415

[17] ICSD#1950, JCPDS-pdf#28-517

[18] ICSD#100799

[19] ICSD#83749

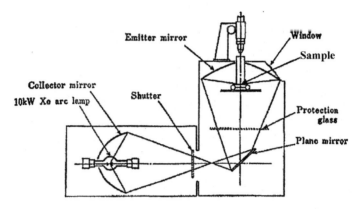

Figure 1. Optical system of the arc-imaging furnace. The light source is Xenon arc lamp. The light is reflected a collection mirror, plane mirror and emittier mirror to heat a sample on a sample stage.

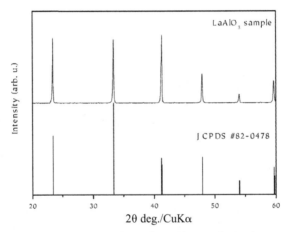

Figure 2. XRD pattern of the melt synthesized $LaAlO_3$ sample. Lattice parameters obtained : a=5.364, c=13.129 are in good agreement with the reference data. (JCPDS #82-0478: a=5.364, c=13.11)

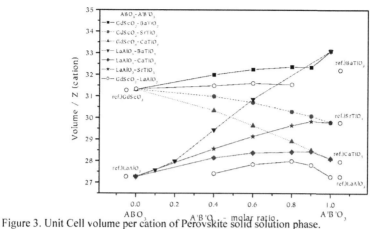

Figure 3. Unit Cell volume per cation of Perovskite solid solution phase.
GdScO$_3$: JCPDS #27-0220, LaAlO$_3$: JCPDS #82-0478, BaTiO$_3$: JCPDS #05-0626, SrTiO$_3$: JCPDS #35-034, CaTiO$_3$: JCPDS #42-0423.

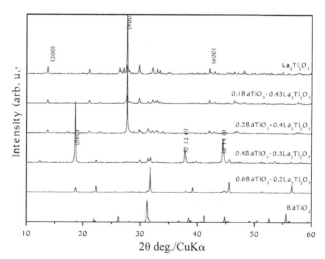

Figure 4. XRD patterns of BaTiO$_3$ – 1/2La$_2$Ti$_2$O$_7$ system. 1/2La$_2$Ti$_2$O$_7$, 0.1BaTiO$_3$-0.45La$_2$Ti$_2$O$_7$, and 0.2BaTiO$_3$-0.4La$_2$Ti$_2$O$_7$ was Pyrochlore. 0.4BaTiO$_3$-0.3La$_2$Ti$_2$O$_7$ was BaLa$_2$Ti$_3$O$_{10}$. 0.6BaTiO$_3$-0.2La$_2$Ti$_2$O$_7$ and BaTiO$_3$ was Perovskite.

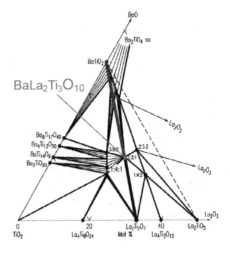

Figure 5. Phase diagram of $BaO-La_2O_3-TiO_2$ system. Skapin et al. [15].

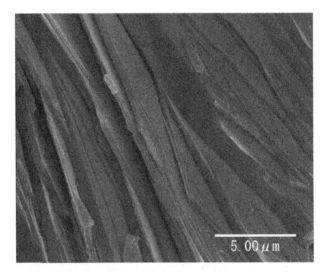

Figure 6. SEM image of melted sample $BaLa_2Ti_3O_{10}$ was crashed. It is observed cleavage surfa

CARBON DERIVED Si₃N₄+SiC MICRO/NANO COMPOSITE

Ján Dusza , Monika Kašiarová, Alexandra Vysocká, Jana Špaková,
Institute of Materials Research, SAS, 043 53 Košice, Slovakia

Miroslav Hnatko, Pavol Šajgalík
Institute of Inorganic Chemistry, SAS, 845 36 Bratislava, Slovakia

ABSTRACT

Silicon nitride/silicon carbide micro-nano-composite was prepared by in situ reaction between carbon and SiO_2 by utilizing carbothermal reduction. The microstructure characteristics and mechanical properties as nano/micro/macro hardness, bending strength, fracture toughness and creep behaviour of a carbon derived Si_3N_4+SiC micro-nano composite have been investigated. The microstructure was studied using SEM, TEM, and HREM techniques. Conventional and depth sensitive methods have been used for characterization of hardness in the load interval from 5 mN to 10 N.

The bending strength was measured in four-point bending, the fracture toughness using single edge V-notched specimens. Creep test was performed in four-point bending using a fixture made of silicon carbide with inner and outer span length of 20 and 40mm, respectively. The measurements were carried out in a creep machine with dead-weight loading system in air at 1350°C with outer fibre stresses in an range from 50 to 150 MPa.

The hardness values of the composite is slightly higher comparing to the hardness of the monolithic material with an evident load-size effect. The strength value of the composite was lower when compared to the monolithic material because of the presence of defects in their microstructure in the form of cluster of large SiC grains and pores. The fractography nearly in all tested specimens revealed fracture origins with approximately size of defects of 30 μm. The composite material exhibits a significantly higher resistance again creep deformation when compared with the monolithic material thanks to the changed chemical composition and viscosity of the intergranular phase and the interlocked silicon nitride grains by the intergranularly located SiC nanoparticle.

INTRODUCTION

Silicon nitride based ceramics exhibit outstanding mechanical and thermo-mechanical properties at high temperatures and are used for many structural applications as parts of energy conversion systems, engines and turbines. During the last decade much effort has been made to optimize the high temperature properties of the silicon nitride based ceramics [1-3] by:

- optimization of the volume fraction and chemical composition of sintering additives,
- crystallisation of the grain boundary phase by heat treatment after densification;
- incorporation of micro- and/or nano- sized secondary refractory phases/particles into the microstructure.

Niihara proposed and applied a ceramic nanocomposite concept and demonstrated that silicon nitride matrix composites containing dispersed nanosized SiC particles exhibit enhanced high temperature strength and creep resistance when compared to monolithic silicon nitride [4]. This improvement is related to a change of grain size/morphology, chemistry of intergranular phase, SiC particles distribution, and structure and chemistry of grain boundaries.

During recent years Si$_3$N$_4$+SiC nanocomposites have been developed using different concepts/processing techniques and characterized as regards their microstructure and room and high temperature mechanical properties [5-7]. Most of the Si$_3$N$_4$+SiC nanocomposites were prepared by hot pressing and the hindered densification due to the presence of SiC particles was overcome using higher densification temperatures in comparison with that for monolithic Si$_3$N$_4$. In case of the gas pressure sintering the densification problem has been solved by using relatively large amount of sintering additives, e.g., mixtures of yttria and alumina or silica. Herrmann et. al.[8] prepared Si$_3$N$_4$+SiC nanocomposite with excellent high temperature properties using gas pressure sintering with Y$_2$O$_3$ additive.

Recent investigations have shown that the microstructure and high-temperature strength of Si$_3$N$_4$ + SiC nano-composites is strongly influenced by the nucleation step of nanoparticles before full densification and that the high-temperature strength was improved only when the SiC nanoparticles were located intergranularly [9].
Pan et al[10] investigated the grain morphology, phase distribution, and the morphology and distribution of SiC particles in Si$_3$N$_4$ + SiC nano-composites by HREM. They used lattice imaging to study the grain boundaries and phase boundaries. They were shown that the thickness of an amorphous film at Si$_3$N$_4$ grain boundaries varies in this material. An amorphous film was found on the Si$_3$N$_4$/SiC boundaries in the case when two crystal parts had a random orientation with respect to each other. Clean phase boundaries were observed in the case when the lattices of two crystal components showed a special orientation relationship. Similar results were found by Dusza et al for SiCN-derived Si$_3$N$_4$ nanocomposite (prepared using SiCN amorphous powder) [11]. Rendtel et al [12] investigated the high temperature properties of a GPS and HP Si$_3$N$_4$+SiC nanoceramics in the range of 1400-1600°C. The annealing of the as–sintered materials promotes crystallization of the intergranular glassy phase resulting in an increase of creep strength and subcritical crack growth resistance.

In the present investigation a carbon-derived Si$_3$N$_4$+SiC nanocomposite was developed with an inexpensive "in situ" method and it's microstructure characteristics, room temperature fracture – mechanical properties and as well as creep behaviour have been investigated.

EXPERIMENTAL MATERIAL AND METHODS
The C-derived Si$_3$N$_4$-SiC nanocomposite with composition listed in Table 1 was used as experimental material. Sample contained additional amount of SiO$_2$ (5.96 wt %) and C (3.62 wt %) with the aim to achieve after densification by carbothermal reduction of SiO$_2$ 5 wt % of SiC.

Table I. Composition of the starting mixture

Compound	Si$_3$N$_4$	Y$_2$O$_3$	C	SiO$_2$
wt %	84.13	4.43	0.43 + 3.62	1.43 + 5.96

The starting mixtures were homogenized in polyethylene bottle with Si$_3$N$_4$ spheres in isopropanol for 24 h. The dried mixture was sieved through 25 μm sieve in order to eliminate the large hard agglomerates. Green discs with a diameter of 48 mm and 5 mm thick were die pressed under the pressure of 30 MPa. Green discs were then embedded into a BN powder bed and positioned into the graphite uniaxial die. Samples were hot-pressed under a specific atmosphere, mechanical pressure, and heating regime at 1750 °C for 2 h.

The microstructure of the nanocomposite was characterized by X-ray difractometry (XRD) and scanning electron microscopy (SEM). XRD analysis was carried out using a Philips X-part difractometer equipped with a CuK$_2$ radiation source. Polished and plasma–etched sections of the bulk materials were examined in SEM.

Specimens with dimensions 3 x 4 x 45 mm were tested in four point bending mode. Specimens were grounded and polished by 15 µm diamond grinding wheel before testing. The two edges on the tensile surface were rounded with a radius around 0.15 mm in order to eliminate a failure from edges of the specimen. The specimens were broken with cross-head speed of 0.5 mm/min, test environment was ambient air. The characteristic strength and Weibull modulus was computed using two - parameter Weibull distribution. The location, type size, and shape of the fracture initiating flaws were examined using SEM and EDX analysis.

Vickers hardness has been measured using indentation loads from 10mN to 10N. For the fracture toughness measurement the sharp notch according to the VAMAS TWA#3/ESIS TC6 Round Robin Instruction using a razor blade [13] was introduced. In all cases the notch tip radius was less than 10 µm. SEM and EDX analyses were used to localize and characterize the fracture origin using standard fractographic methods [14, 15].

Creep tests were performed in four-point bending using a fixture made of silicon carbide with the inner and outer span length of 20 and 40mm, respectively. The measurements were carried out in a creep machine with a dead-weight loading system in air, at temperatures between 1200°C and 1450°C with outer fibre stresses in the range from 50 to 150 MPa. The sample deflection was continuously recorded during the creep test. From the deflection data, the outer fibre strain was calculated as a function of time, t, by the method of Hollenberg et al [16] and taken as the creep strain, ε. The creep rate was calculated from the slope of the ε versus t curve. The steady–state creep rate is usually described by the Norton equation :

$$\dot{\varepsilon} = A\sigma^n \frac{1}{d^m} \exp\left(-\frac{Q_c}{RT}\right) \tag{1}$$

where A is a constant, depending on the respective material properties and microstructure, σ is the stress, n is the stress exponent, d is the grain size, m is the grain size exponent, Q_C is the activation energy of creep, and T and R have their usual meanings.

EXPERIMENTAL RESULTS AND DISCUSSION

Characteristic microstructure of the experimental material is illustrated in Fig. 1. The Si$_3$N$_4$ - SiC nanocomposite consists of a very fine, homogeneously distributed Si$_3$N$_4$ grains with a low aspect ratio. The composite additionally contains globular nano and submicron-sized SiC particles located intergranularly in the Si$_3$N$_4$ grains (average particle size approximately 453 nm), or intragranularly between the Si$_3$N$_4$ grains (average particle size approximately 250 nm).

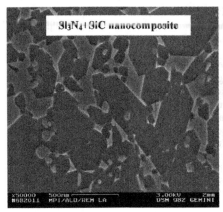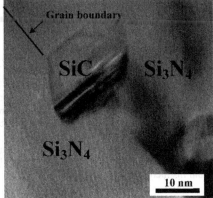

Fig. 1 Characteristic microstructure of the carbon - derived nano-composite, a - SEM - plasma etched, b – TEM. The intergranularly located SiC particles are not visible in Fig. 1a and appear as an intergranular phase but act as a part of interconnection between Si$_3$N$_4$ grains, Fig.1b.

It was difficult to distinguish between the SiC particles located integranularly and the grain boundary phase because they are affected by plasma etching similarly. The average Si$_3$N$_4$ grain size (diameter) is 140 nm and grains with a diameter larger than 500 nm occur in the microstructure occasionally, only. The volume fraction of the SiC nanoparticles can be estimated approximately as 5 volume percent. X-ray analysis revealed that the main phase in the material is β-Si$_3$N$_4$ with a low amount of α-Si$_3$N$_4$ and β-SiC. Beside the Si$_3$N$_4$ and SiC, some additional crystalline phases have been detected in the composite mainly YSiO$_2$N and Y$_2$Si$_3$O$_3$N$_4$.

According to the results, the hardness of the micro/nano composite is slightly higher when compared with the hardness values of the monolithic material. The Weibull distribution of the measured four point flexure strength values of the investigated nanocomposite is showed in Fig. 2. The two-parameter Weibull statistics results in the characteristic strength σ$_0$ = 675 MPa and Weibull modulus m = 6.4.

Fractography revealed that the location of fracture origins is in most cases in the volume of specimens (48 %), but also near-surface (20 %), surface (20 %) and edge located origins (12 %), were present. The study of the size and shape of the fracture origins revealed that their dimensions are in the range from 10 μm to 180 μm with the majority of flaws in the range from 20 to 30 μm. Their shape is in the most cases elliptical, but circular shaped fracture origins have been occurred, too. Fractographic analysis of the fracture surfaces of the failed specimens combined with EDX analysis revealed two types of technological defects acting in the studied material as a failure initiating flaw. The first type was agglomerate of SiC grains, often connected with a porous area, Fig. 3a. The second type of fracture origin was a cluster of porous area, Fig.3b.

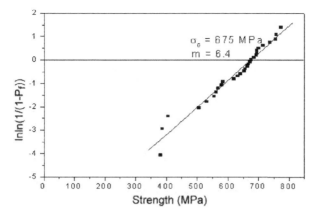

Fig. 2 The Weibull distribution of the measured four point flexure strength values

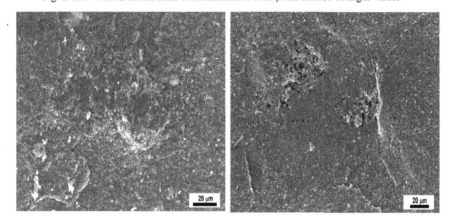

Fig. 3 Typical flaws in Si_3N_4-SiC nanocomposite, a) cluster of large SiC grains,
b) porosity areas

It was possible to find a fracture origin almost in all fractured specimens, only in a few cases the origin was lost by impact between the test-piece and the testing jig.

Fracture toughness measured using the single edge V-notch beam method was found to have a value of 3.75 ± 0.53 MPam$^{1/2}$. This value is significantly lower comparing to the value measured for a similar material using indentation technique[17]. There is an other way to estimate the fracture toughness of the investigated material using the Griffith relationship and results of the fractographic examinations concerning the size and shape of defects.

The value of fracture toughness estimated from this relationship was found as 4.4 MPam$^{1/2}$. This value is slightly higher comparing to the measured value using SEVNB technique. The reason of such a difference can be caused by difficulties in the estimation of the Y value for individual defects but also by the fact that the origins have different composition/structure that the surrounding area. These origins mainly consist of a large SiC grains, which can cause a higher toughening effect than finer grains of Si₃N₄-SiC matrix. Also a porous region as fracture origin can caused a higher value of fracture toughness.

The creep deformation of both, monolithic and composite ceramics can be characterized by a primary creep range and a pronounced steady-state creep range. Ternary creep regime was not observed in the studied materials at the applied test conditions. The composite ceramics exhibits only minimum creep deformation up to 1300 °C. Significant creep deformation was detected only at the temperatures from 1350 °C to 1450°C. In spite of the lower grain diameter of Si₃N₄ grains in the nanocomposite, the creep resistance of the nanocomposite was found to be significantly higher when compared to the creep resistance of the monolithic Si₃N₄. At the temperature of 1350 °C the composite exhibits similar creep strain at 150 MPa/150h as the monolithic material at 50MPa/25h (Fig. 4).

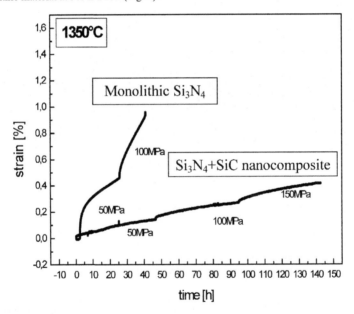

Fig. 4 Comparison of the creep deformation of monolithic silicon nitride and of the C-derived nanocomposite

The stress exponents of the nanocomposite are in the interval from 0.8 to 1.28 and the apparent activation energy is 480 kJ/mol. The stress exponents of the monolithic material are in the interval from 1.0 to 2.0 with the apparent activation energy of 372 kJ/mol. TEM observation of the crept specimens revealed no cavitation in the nanocomposite over the studied temperature/stress interval, and only a limited cavitation in the monolithic material at and above 1400 °C.

It seems that the intergranularly located SiC nanoparticles are very important. They hinder the grain growth of the Si$_3$N$_4$, changing the shape and chemistry of the grain boundaries and grain boundary phases. What's more important, the intergranularly located SiC nanoparticles interlock two neighbouring Si$_3$N$_4$ grains with a " puzzle" character.

Pezzotti et al[18] found that the anion structure of SiO$_2$ glasses segregated at grain boundaries in Si$_3$N$_4$ and SiC ceramics is affected by the presence of dissolved ions which is, in the case of SiC materials, approximately 15 at%. Glasses containing C substituting oxygen in the (SiO)$^{4-}$ tetrahedral network could result in a higher bonding density per unit volume of the glass and in the increased glass viscosity. In the investigated nanocomposite, the presence of the carbon and SiC probably leads to a similar effect, and the remained glass exhibits high viscosity. The detailed study of the composition of intergranular phase is the subject of future investigation.

The stress exponents and activation energies together with the results of TEM examination show that there probably exist different creep mechanisms in the monolithic silicon nitride and in the nanocomposite; namely cavitation together with grain boundary sliding and solution/precipitation, respectively.

CONSLUSION

- Si$_3$N$_4$ +SiC nanocomposite processed by an inexpensive in-situ method utilizing carbothermal reduction of C+ SiO$_2$ during sintering exhibits nearly defect free, very fine microstructures with inter and intragranulary located SiC nanoparticles;
- the characterlstlc strength of the investigated nanocomposite was σ_0 = 675 MPa with the Weibull modulus of 6.4. In almost all specimens processing related fracture origin was found using fractographic methods with a dimension in a range from 10 µm to 180 µm. Two types of technological flaws were identified; agglomerates of large SiC grains and agglomerates of porosity areas;
- the intergranular SiC nanoparticles hinder the Si$_3$N$_4$ grain growth, interlock the neighbouring Si$_3$N$_4$ grains and change the volume fraction, geometry, and chemical composition of the grain boundary phase. This all leads to a change of the creep mechanisms and to a significantly improved creep resistance of the Si$_3$N$_4$ +SiC nanocomposite.

ACKNOWLEDGEMENT

This work was realized with the financial support of the Slovak Grant Agency, under the contract No. 2/4173/04, of the Royal Society, of NANOSMART, Centre of Excellence of SAS, and by 2003 SO 51/03R8 06 00/03R 06 03-2003. The authors acknowledge the help of G. Roebben and H. Labitzke.

REFERENCES

[1] M. J. Hoffmann, "High Temperature Properties of Yb Containing Si$_3$N$_4$," In *Tailoring of Mechanical Properties of Si$_3$N$_4$ Ceramics*, ed. M. J. Hoffmann and G. Petzow, Kluwer, Dordrecht, 233-244 (1993)

[2] M. K. Cinibulk, G.Thomas, and S. M. Johnson, "Grain- Boundary – Phase Crystallization and Strength of Silicon Nitride Sintered with YSiAlON Glass," *J. Am. Ceram. Soc.*, **73**, 1606-1612 (1990)

[3] F. F. Lange, "Effect of Microstructure on Strength of Si$_3$N$_4$-SiC Composite System," *J. Am. Ceram. Soc.*, **56**, 445-450 (1973)

[4] K. Niihara, "New Design Concept of Structural Ceramics-Ceramic Nanocomposites", *J. Ceram. Soc. Jpn.*, **99**, 974-982 (1991)

[5] A. Rendtel, H. Hübner, M. Herrmann, Ch. Shubert, "Si$_3$N$_4$/SiC Nanocomposite Materials": II, Hot Strength, Creep, and Oxidation Resistance," *J. Am. Ceram. Soc.*, **81**, 1109-1120 (1998)

[6] H. Klemm, M. Herrmann, and Ch. Schubert, "High–Temperature Properties of Si$_3$N$_4$/SiC Nanocomposites," *Ceram. Eng. Sci. Proc.*, **21**, 3-713-720 (2000)

[7] T. Rouxel, F. Wakai, S. Sakguchi, "R-curve Behavior and Stable Crack Growth at Elevated Temperatures (1500°-1650°C) in a Si$_3$N$_4$/SiCN naocomposite," *J. Am. Ceram. Soc.*, **77**, 3237-3243 (1994)

[8] M. Herrmann, Ch. Schubert, A. Rendtel, and H. Hübner, "Silicon Nitride/Silicon Carbide Nanocomposites Materials": I. Fabrication and Mechanical Properties at Room Temperature," *J. Am. Ceram. Soc.*, **81**, 1094-1108 (1998)

[9] H. Park, H. Kim, and K. Niihara, "Microstructure and High-temperature Strength of Si$_3$N$_4$ -SiC Nanocomposite," *J. Eur. Ceram. Soc.*, **18**, 907-914 (1988)

[10] X.Pan, J.Mayer and M. Rühle, "Silicon Nitride Based Nanocomposite," *J. Am. Ceram. Soc.*, **79**, 585-90 (1996)

[11] J. Dusza, P. Šajgalík, M. Steen and E. Semerad, "Low-Cycle Fatigue Strength under Step Loading of a Si$_3$N$_4$+SiC Nanocomposite at 1350 C, " *Journal of Materials Science*, **36**, 4469-4477 (2001)

[12] A. Rendtel and H. Hübner, "Effect of Heat Treatment on Microstructure and Creep behaviour of silicon nitride based ceramics, "*J. Europ. Ceram. Soc.*, **22**, 2517 – 2525 (2002)

[13] J. Kübler, "Fracture Toughness of Ceramics Using the SEVNB Methods: Preliminary Results," *Ceram. Eng. Sci. Proc.*, **18**, 155-62 (1997).

[14] G. D. Quinn and J. J. Swab, "Fractography and Estimates of Fracture Origin Size from Fracture Mechanics," *Ceram. Eng. and Sci. Proc.*, **17**, 51-58 (1996).

[15] J. Dusza, M. Steen, "Fractography and fracture mechanics property assessment of advanced structural ceramics, *Internat. Mater. Reviews*, **44** (1999)

[16] G.W. Hollemberg, G.R.Terwilliger, and R.S.Gordon, "Calculation of stresses and strains in four-point bending creep tests", *J. Am. Ceram. Soc.*, **54**, 196-199 (1971).

[17] J. Dusza and P. Šajgalík, "Fracture Characteristics of Layered and Nano-particle reinforced Si$_3$N$_4$," In *Advanced multilayered and fibre feinforced composites*, ed. Y. M. Haddad, Dordrecht, Kluwer Academic Publishers, 187-205 (1998)

[18] G. Pezzotti, T. Wakasugi, T. Nishida, R. Ota, H.-J. Kleebe and K. Ota, "Chemistry and inherent viscosity of glasses segregated at grain boundaries of silicon nitride and silicon carbide ceramics", *J. Non-Cryst. Solids.*, **271**, 79-87 (2000)

Dispersion

MODIFICATION OF NANOSIZE SILICA PARTICLE SURFACES TO IMPROVE DISPERSION IN A POLYMER MATRIX

Chika Takai, Masayoshi Fuji, and Minoru Takahashi
Ceramics Research Laboratory, Nagoya Institute of Technology
Tajimi 507-0071, Japan

ABSTRACT

Composite materials composed of inorganic particles and polymers have been developed with the purpose of improving the physical properties, such as strength, heat resistance, and elastic modulus. There is no repulsive force between particle surfaces in an organic medium, hence, the interaction between particles is always attractive. Because of this, it is difficult to uniformly disperse inorganic particles in polymers. To achieve good dispersion in a polymer, it is necessary to modify the surface of inorganic particles. In this paper, we used chemical surface modification to change the surface design of inorganic particles. To uniformly disperse particles in a polymer we used organic compounds with a chemical structure similar to the polymer to chemically modify surfaces and change the molecular level particle surface design.

INTRODUCTION

In the fabrication of organic/inorganic composite, improvements in processing, cost reductions, and improvements in physical property can be expected with increasing particle loading in a polymer matrix. [1, 2] In order to increase the leading efficiency, the interfacial energy between the particles and the polymer matrix should be considered as well as the shapesize, and surface of the particles. Recently, with the development of nanotechnology, particle size has become to be controlled on a nano scale. Relative to larger particles, nanosize particles are known to have different electrical, optical, magnetical, and chemical properties. However, nanosize particles aggregate sassily owing to their specific surface area. Therefore, it is difficult to uniformly disperse nanoparticles into a matrix using only mechanical mixing. [3] This is due to the interfacial interaction between the particle surface and the matrix during the mixing process.

One way to improve the wettability of particles by the matrix is through chemical surface modification using a coupling reagent. [3] It has been reported a polymer coating on the particles can prevent aggregation. [4]

In order to decrease the difference in interfacial energy between the particles and the matrix, a novel molecular level surface treatment was proposed. It is expected that this surface modification will prevent particle aggregation. In this study, polyimide was used as the polymer matrix and its monomers were used as reagents to modify particle surfaces. Surface characterization of the treated particles was performed using simultaneous thermogravimetric and differential thermal analysis (TG-DTA), and Fourier Transform Infrared spectrometry (FT-IR). Additionally, an improvement in the dispersability in N-methyl-2-pyrrolidone (NMP), which is a solvent for the polyimide precursor, was confirmed.

EXPERIMENTAL SECTION
Materials

Fine silica powder (OX50, specific surface area 50 m^2/g, particle size 40 nm) and Aminopropyl triethoxysilane (APTS) were purchased from Nippon Aerosil Co., and Shin-Etsu Chemical Co., Ltd., respectively. N-methyl-2-pyrrolidone (NMP) was obtained from Aldrich. Hexane, Pyromellitic dianhydride (PMDA), and 4, 4'-Diaminodiphenylether (DDE) were purchased from Wako Pure Chemical Industries Ltd., Osaka, Japan.

Modification of silica with APTS

Chemical surface modification was accomplished through is a reaction of APTS with the hydroxyl groups of the silica substrate. This reaction was carried out using the autoclave method[5, 6] at 235 □ and under 30 atm for 1 h. The degree of surface modification was controlled using a modifier content of 3 –OH/nm^2 surface density of the hydroxyl groups determined by the Grignard reagent method.[7] Before the measurement, the sample was degassed at 200 □ for 4 h.

Synthesis of surface treated samples

The surface treatment was performed as follows: Modified silica with APTS, NH$_2$ modified silica, was suspended in NMP, and then PMDA was introduced into the suspension. The one functional anhydride group of PMDA reacts with the surface amino group. Then, DDE was dissolved into the suspension to react with the other groups of PMDA. After repeating these two reactions, a surface treated sample (i.e., coated particle) was obtained.

Determination of the surface density of the modifier

The surface density of the modifier was estimated from the number of modifiers and the specific surface area. The number of aminopropyl groups was determined using TG/DTA (Thermo Plus TG8120, Rigaku Co., Japan) as shown in Figure 1.[8] The measurement was carried out under oxygen at a flowing at 250 ml/min. The surface aminopropyl group density, d_A, was calculated using Eq. (1),

$$d_A[/\text{nm}^2] = \frac{(\Delta w_{CH} - \Delta w_{OH})N_A}{M_W S_{N2}} \times 10^{-18} \qquad (1)$$

where Δw_{CH} and Δw_{OH} are the ratio of weight loss for the modified and unmodified samples measured by thermogravimetry, M_w is the molar weight of the aminopropyl group, S_{N2} is the BET specific surface area measured from the nitrogen adsorption isotherm and N_A is Avogadro's number.

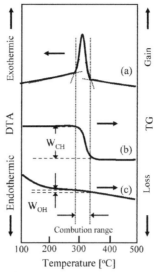

Figure 1. Estimation of the number of aminopropyl groups introduced onto the silica surface from simultaneous thermogravimetric analysis (TG) and differential thermal analysis (DTA); (a) DTA curve for modified silica, (b) TGA curve for modified silica, (c) TGA curve for unmodified silica.

Characterization

The FT-IR study was carried out using a FT-IR 6200 (JASCO Co., Japan) to characterize the particle surface. [9] The amount of coating on the particle surface was determined from TG/DTA analysis. Dispersability of these particles into NMP was confirmed by visual observation.

RESULTS AND DISCUSSION

The surface modification of the silica powder with APTS was successfully completed. The relationships between modifier content and specific surface area and surface density of the modification groups are shown in Figure 2. The degree of modification is the fraction of converted hydroxyl groups relative to all of the hydroxyl groups. As shown in Fig. 2, the density of the surface modifier groups increased significantly with increasing modifier content. The highest aminopropyl group density was 3.32 $-OR/nm^2$ with a modifier content of 0.58 ml/g. This indicates that all of the surface hydroxyl groups reacted with APTS. The reason why the surface density of modifier is higher than that of hydroxyl groups is believed to be due to a reaction with the APTS.

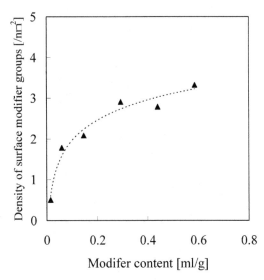

Figure 2. The relationships between modifier content and the density of the surface modification groups.

Figure 3 shows the IR spectra for calcined silica, NH_2 modified silica, and coated silica. For the calcined silica, a band at 3747 cm^{-1} due to the OH stretching vibration of the free silanol group was observed. In NH_2 modified silica, this band disappeared and three new bands between 2915 and 2985 cm^{-1} appeared. These bands are due to the C-H stretching vibration, indicating the attachment of APTS to the silica. The IR spectra of the resultant coated silica contain peaks at 1456, 1740, and 1777. These bands are caused by C-N stretching, asymmetric imide C=O stretching, and symmetric C=O stretching, respectively. Hence, the NH_2 functional groups in NH_2 modified silica are able to react with the anhydride end groups of PMDA and the other end groups of PMDA are able to react with the end groups of DDE.

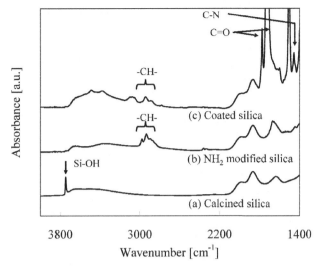

Figure 3. IR spectra of (a) calcined silica, (b) NH$_2$ modified silica, and (c) coated silica.

In order to assess the dispersability of the modified and the coated silica particles in NMP, a sedimentation test was performed. This was a simple test in which that the silica sample was introduced into NMP and stirred. The dispersability was assessed qualitatively by visual observation. The calcined silica powder settled immediately. On the other hand, no sedimentation was observed for the NH$_2$ modified silica of the coated silica powder.

The amount of organic compound on the particle surface relative to the particle weight was obtained from TG/DTA analysis. For the coated silica, it was 6.91 %. From this result, the thickness of organic compound on the particle surface was calculated to be about 0.45 nm.

To determine the minimum coating thickness necessary to disperse the primary particles, the van der Waals force between the particles was calculated. Van der Waals force V$_A$ was determined from eq. (2),[10]

$$V_A = -\frac{Aa}{12H} \qquad (2)$$

where A, a, and H are the Hamaker constant, the particle size, and the distance between particles, respectively. The Hamaker constant is unique for a given material and can be calculated using eq. (3),[11,12]

$$A_{12} = \pi^2 q_1 q_2 \beta_{12} \qquad (3)$$

where q is the number of atoms per unit volume, and β is the constant in London's equation for the interaction between two atoms. Approximations of β, have been made by various authors. In this report, an approximation of β according to London[13] was employed. Figure 4 shows the van der Waals force as a function of distance between particles. With increasing distance larger than 2.87 nm, the van der Waals attractive force increases rapidly. Therefore, a coating thickness that is more than 2.87 nm should prevent particle aggregation. From this estimation, the thickness of the polymer coating on the particle in this experiment proved to be too thin to obtain primary particle dispersion. Further research could improve particle dispersability.

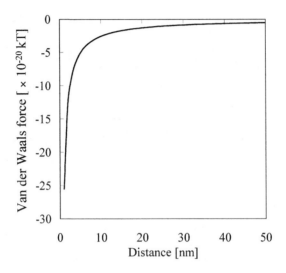

Figure 4. Van der Waals force as a function of distance between particles calculated from equation (2) and (3).

CONCLUSION

In order to improve particle dispersability, a novel molecular level surface treatment was attempted. The surface density of the modifier can be controlled by modifier content. FT-IR and TG/DTA analysis confirmed that the surface was successfully modified.

The dispersability of the coated particles in NMP was remarkably advanced using this technique. The thickness of the polymer coating was estimated to be about 0.45 nm from TG/DTA analysis. This is less than the 2.87 nm thickness that was calculated to be enough for primary particle dispersion. However, further investigation could achieve the thickness.

This technique has prospects for further improvement of particle dispersability in organic/inorganic nanocomposite.

REFERENCES

[1]B. K. Larson, L. T. Drazal, "Interphase Development in a Liquid Composite Molding Environment," *Proceeding of the American Society of Composites, Technical Conference,* 7th, 286-293, (1992)

[2]K. W. Thomson, L. J. Broutman, "The effect of water on the fracture surface energy of fiber-reinforced composite materials," *Polymer Composites,* **3[3]**, 113-117, (1982)

[3]K. Matsumoto, T. Yoshida, K. Okuyama, "Dispersion of surface modified silica nanoparticles into a resin by using a twin screw extruder," *Funtai Kogaku Kaishi*, 40(7), 489-496, (2003)

[4]K. Ohno, K. Koh, Y. Ysujii, T. Fukuda," Synthesis of Gold Nanoparticles Coated with Well-Defined, High-Density Polymer Brushes by Surface-Initiated Living Radical Polymerization," *Macromolecules*, **35**, 8989-8993, (2002)

[5]H. Utsugi, H. Horikoshi, T. Matsuzawa, "Mechanism of esterification of alcohols with surface silanols and hydrolysis of surface esters on silica gels," *J. Colloid Interface Sci.*, **50**, 154-161, (1975)

[6]Lange, K. R, "Reactive organic molecules on silica surfaces," *Chem. Ind.*, 14, 441-3, (1968)

[7]T. Kanazawa, M. Chikazawa, T. Takei, and K. Mukasa, "Characterization of surface OH groups on porous glass," *Yogyo-Kyokai-Shi,* **92[11]**, 654-659, (1984)

[8]M. Fuji, S. Ueno, T. Takei, T. Watanabe, M. Chikazawa, "Surface structural analysis of fine silica powder modified with butyl alcohol," *Colloid Polym. Sci.*, **278**, 30-36, (2000)

[9]L. Chyi-Ming, W. Zhen-Wei, and W. Kung-Hwa, "Synthesis and Properties of Covalently Bonded Layered Silicates/Polyimide (BTDA-ODA) Nanocomposites," *Chem. Mater.*, **14**, 3016-3021, (2002)

[10]J. Gregory, "The calculation of Hamaker constants ," *Adv. Colloid Interface Sci.*, **2[4]**, 396-417, (1970)

[11]J. Visser, "On Hamaker constants: A comparison between Hamaker constants and Lifshitz-van der Waals constants," *Adv. Colloid Interface Sci.*, **3[4]**, 331-363, (1972)

[12]H. C. Hamaker, "The London-van der Waals attraction between spherical particles," *Physica (The Hague)*, **4**, 1058-1072, (1937)

[13]R. Eisenschitz, F. London,"The relation between the van der Waals force and the homopolar valence forces," *Z. Phys.*, **60**, 491-527, (1930)

Acknowledgements

A part of this study has been supported by The Tatematsu foundation and Research Foundation for the Electrotechnology of Chubu (REFEC).

POSSIBILITY OF COMB-GRAFT COPOLYMERS AS DISPERSANTS FOR SIC
SUSPENSIONS IN ETHANOL

Toshio KAKUI, Mitsuru ISHII
Chemicals Research Laboratories, Chemicals Division, Lion Corporation
13-12, Hirai 7-chome, Edogawa-ku
Tokyo, 132-0035, Japan

Hidehiro KAMIYA
Department Chemical Engineering, BASE, Tokyo University of Agriculture and Technology
24-16, Nakamachi 2-chome, Koganei-shi
Tokyo, 184-8588, Japan

ABSTRACT

In our previous studies on linear and branched polyethyleneimines (PEIs) and comb-graft
PEIs, the effect of a polymer dispersant's molecular structure on ethanol suspension with SiC,
Si_3N_4 and Al_2O_3 was examined. The steric repulsion force between particles by the branched
chains of a PEI, which reduced the ethanol suspension viscosity, was characterized using a
colloidal probe AFM. The branched chains of a polymer dispersant affect the steric repulsion and
the suspension viscosity. However, the effect of the comb-graft copolymer on ceramic
suspensions has not been investigated. In this study, using comb-graft copolymers from five
types of backbone compounds and different molecular weight polypropylene glycol branches,
the effects of the molecular structure of a comb-graft copolymer as a dispersant in concentrated
SiC/ethanol suspensions was investigated. High molecular weight backbone copolymers , such
as PEI, hydroxypropyl cellulose, and aqueous nylon, adsorbed on SiC particles and reduced the
suspension viscosity. It was easy to control the rheological properties of the suspension, because
the viscosity is gradually reduced with increasing amount of copolymer. The effects of the
backbone and branches in a comb-graft copolymer, and the amount of copolymer adsorbed on a
SiC particle surface are discussed, and the mechanism for the effect of the copolymer on
suspension viscosity is discussed.

INTRODUCTION

There are many applications for ceramic powders in new advanced materials, e.g., in high
strength structural materials, electronic and optical materials, catalyst carriers, etc. To
manufacture ceramic products with a high performance and reliability at a low cost using
forming processes such as tape casting, slip casting, and spray drying, fine ceramic powders have
to be handled as a homogeneous suspension with a high solid volume fraction and adequate
fluidity. Nonaqueous suspensions are often preferred when multilayer ceramic substrates are
prepared by tape casting, where the presence of water causes a deleterious effect.

Various types of dispersants have been proposed to control the stability and rheological
properties of a suspension. Adsorbed dispersants on a particle surface can result in electrostatic
and steric repulsive forces between particles that promote the dispersiblity of the particles. The
electrostatic repulsion between particles in various organic solvent suspensions with alkyl
amines has been discussed[1,2]. Some researchers have reported that steric repulsion between
particles facilitates dispersibility in a nonaqueous suspension. Surfactants[3,4] with various

functional groups, and a few polymer dispersants[5, 6, 7] for nonaqueous dense suspensions have been proposed for various ceramic powders. However, the effect of the molecular structure of these dispersants on dispersability has not been determined.

In our previous studies[8,9,10], we investigated the effect of the polymer dispersant's molecular structure on ceramic suspensions with good dispersibility in aqueous and nonaqueous systems. In dense aqueous suspensions with Al_2O_3, SiC, and Si_3N_4, the optimum molecular structure of the anionic acrylate copolymer with different hydrophilic to hydrophobic group ratios was determined based on the microscopic surface interaction by an atomic force microscope (AFM), and the macroscopic behavior of the suspension (i.e., its viscosity and aggregate size distribution). Additionally, the effect of polyethyleneimine (PEI) and its derivatives with different structures, (i.e., linear and branched PEIs, and comb-graft PEIs), on dense ethanol suspensions with SiC, Si_3N_4 and Al_2O_3 has recently been examined[11]. The steric repulsion force between particles by branched chains of a PEI was characterized using a colloidal probe AFM. The branched chains of a polymer dispersant play an important role in the steric repulsion and the suspension viscosity [12].

Comb-graft copolymers with a backbone (main chain) and some branches (side chains), as is shown in Figure 1, have been investigated as dispersants to improve the steric repulsion of a polymer. Some researchers discussed the effect of the backbone and branches in a comb-graft polymer in different fields. Comb polymers with polyacrylic acid, a polyelectrolyte backbone, and charge neutral polyethylene oxide teeth (i.e., branches), drastically improved the stability of concentrated cement suspensions. Effects of the comb polymer architecture on the rheological property evolution were systematically investigated[13]. Using sodium polyacrylate with methoxy polyethylene glycol branches, the rheological behavior of a concentrated aqueous $CaCO_3$ suspension was examined, and the depletion effect, which also has an important effect on suspension viscosity, was discussed[14]. As a dispersant for a coal water mixture (CWM) with a high solid fraction and high fluidity, hydrophilic sodium polyacrylate with hydrophobic polystyrene branches was also examined[15]. Various nonionic comb-graft copolymers were prepared from hydrophilic polyethylene glycol branches and some backbones, such as hydrophobic poly-4-methylstyrene and polynonylphenol, and hydrophilic polyethyleneimine [16,17,18]. The optimum molecular structure of the nonionic polymer dispersants was discussed based on the stability and dynamic properties of the CWMs. However, the effect of a comb-graft copolymer on ceramic suspensions has not been investigated.

The current study focuses on the effect of the molecular structure of comb-graft copolymers on a SiC ethanol suspension. By using a series of comb-graft copolymers containing polypropylene glycol (PPG) branches with different lengths prepared from five types of backbone compounds as a dispersant, the roles of a backbone and the side branches (i.e., that are the teeth of the comb), on a SiC suspension were investigated. The optimum backbone and the effect of the branch's length were evaluated. The

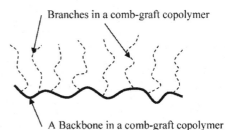

Branches in a comb-graft copolymer

A Backbone in a comb-graft copolymer

Figure 1. A model structure of a comb-graft copolymer

mechanism of dispersion for an SiC ethanol suspension with comb-graft copolymers will be discussed based on the amount of polymer adsorbed onto the SiC particles. Preparation of Comb-graft Copolymers with Different Molecular Structures

A series of comb-graft copolymers was prepared from backbone compounds with different molecular structures and polypropylene glycol (PPG) branches with different molecular weights. Table I shows the characteristics of five kinds of backbone compounds in the comb-graft copolymer. The compounds have molecular weights ranging from 60 to about 40,000, and functional groups with active hydrogen such as -NHn, -OH, and –NHCO-. Three kinds of polyethyleneimine (PEI) (Nippon Syokubai Co., Ltd., Japan), hydroxypropyl cellulose (HPC-SL, Nippon Soda Co., Ltd., Japan), and aqueous nylon (AQ-Nylon, Toray Co., Ltd., Japan) were used as polymer backbone compounds with lots of active hydorogen. Sorbitol and ethylenediamine (Junsei Chemical Co., Ltd., Japan), with low molecular weight (Mw) and a little active hydrogen, were also used.

Table I. The characteristics of five backbone compounds in the comb-graft copolymer

Molecular structure	Sample Name	Molecular weight	Number [1] of active hydrogen
$_2HN-[CH_2CH_2N]_m-[CH_2CH_2N]_n-CH_2CH_2NH_2$ CH$_2$ ... CH$_2$ CH$_2$ ... CH$_2$ NH(CH$_2$CH$_2$NH)x-H N-(CH$_2$CH$_2$NH)y-H (CH$_2$CH$_2$NH)z-H	PEI-003	300	9
	PEI-012	1,200	29
	PEI-018	1,800	44
Polyethyleneimine	PEI-100	10,000	244
(-NHCO-R'-NHCO-R-) $_n$ Aqueous nylon	AQNY	~ 25,000	~ 105
[Hydroxypropyl cellulose structure] R=H, CH$_2$CH(CH$_3$)OH Hydroxypropyl cellulose	HPC	~ 40,000	~ 187
CH$_2$-CH-CH-CH-CH-CH$_2$ OH OH OH OH OH OH	Sorbitol	182	6
H$_2$N-CH$_2$-CH$_2$-NH$_2$	Ethylenediamine	60	4

1) These were calculated from the molecular structure and the functional group information of the compounds.

The number of active hydrogen in these compounds is shown in Table I. This number will be similar to the number of branches in each comb-graft copolymer, because the PPG branches are prepared by adducting propylene oxide into each active hydrogen in a backbone compound using KOH catalysis[9].

As the representative reaction for the synthesis of the comb-graft copolymer with PPG branches, the PEI comb-graft copolymer is illustrated in Scheme 1. 1 mole of propylene oxide is first adducted into each active hydrogen in the amines in the PEI without the catalysis. This compound is the comb-graft copolymer with branches of 1 mole propylene glycol. Next, the PEI copolymer containing PPGs with some molecular weight was synthesized by incorporating an appropriate amount of propylene oxide into the above PEI copolymer using KOH catalysis. The molecular weight of PPG on one branch was determined by a calculation based on the number of active hydrogen, and the actual total amount of reacted propylene oxide. The length (i.e., molecular weight (Mw)) is controlled by the amount of adducted propylene oxides. The molecular weight of each PPG branch in the comb-graft copolymers was designed to range from 60 to about 4,300.

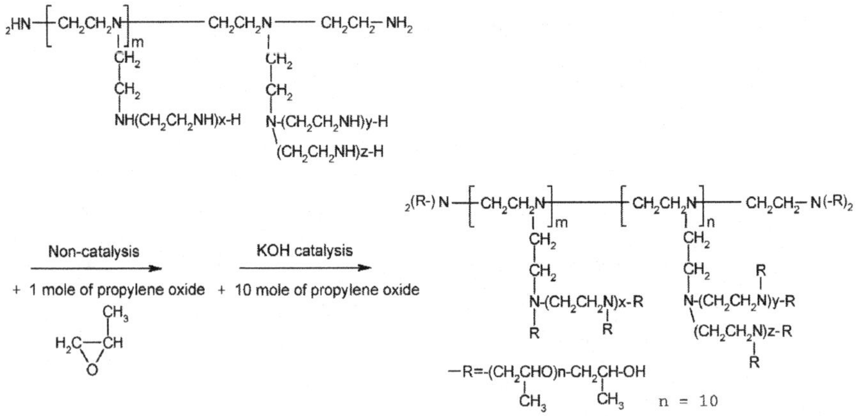

Scheme 1. The synthesis of the PEI comb-graft copolymer with PPG branches as a representative reaction

Preparation and Characterization of Dense Ethanol Suspensions

Three kinds of ceramic powders, SiC and Si_3N_4 (A-1 and NU-1, Showa Denko K.K., Japan), and α-Al_2O_3 (AL-160SG, Nippon Light Metal Company, Ltd., Japan), were used in this study. Table II shows the representative characteristics of each fine ceramic powder. Ethanol (HPLC grade, Junsei Chemical Co., Ltd., Japan) was used without removing any trace of water or further purification. The volume fraction of SiC powder in the ethanol suspensions ranged from 16 to 65 vol.%. The powder was mixed in ethanol with different amounts of polymer dispersant ranging from 0.14 to 0.71 mg/m^2, and then ball-milled for 24 hrs. The apparent viscosity of the ethanol suspension was determined by a BL type viscometer at shear rates of 0.63, 1.29, 3.14, and 6.28s^{-1}, because the apparent viscosity can easily reveal macroscopic behavior due to aggregation and dispersion.

Table II. Representative characteristics of the fine ceramics powders studied

	Average particles size	Specific surface area	True specific gravity	Isoelectric point
SiC	0.4μm	14.0 m^2/g	3.20 g/cm^3	3-5
Si$_3$N$_4$	1.5	8.5	3.20	5-6
Al$_2$O$_3$	0.6	7.0	3.93	8.5-9.5

The amount of polymer dispersant adsorbed on the particles in each 20 vol.% ethanol suspension was calculated from the difference between the original concentration of the polymer in ethanol and that of the free polymer in the supernatant. The dispersant concentration ranged from 0.14 to 0.71 mg/m^2, based on adsorption isotherms. In the other cases, the dispersant concentration was fixed at 0.71 mg/m^2. The supernatants containing the free polymer were removed from each suspension using centrifugal sedimentation at 5,000 rpm for 30 min. The amount of free polymer was determined by removing ethanol from the supernatant through distillation and drying with a vacuum dryer at 80°C for 1hr. The eluted amount of the organic compounds from the SiC powder, and the amount of polymer adsorption in the ball mill were measured and corrected as a blank value.

RESULTS

Effects of Backbone Compounds in Comb-graft Copolymer

The effects of the backbones of the comb-graft copolymer on SiC ethanol suspension characteristics were investigated . Various copolymers with PPG branches were prepared from five types of backbone compounds with different molecular structures and weights. The molecular weight (Mw) of one PPG branch was engineered at about 700, although the results were Mws ranging from 580 to 930. Table III shows the viscosity of 20 vol.% SiC ethanol suspensions each containing a copolymer concentration fixed at 0.71 mg/m^2, which was a sufficient amount for good dispersibility[11]. As is shown in previous work, the original PEIs with a Mw of 1,200 and 10,000 reduced the viscosity down to about 10 mPa s. However, when the other original backbone compounds were used as a dispersant, higher fluidity SiC ethanol suspensions (i.e., relative to a blank) were not obtained. Comb-graft copolymers with many branches from polymers such as PEI and AQNY reduced the suspension viscosity, but the copolymer from HPC did not. Copolymers with little branches from compounds such as sorbitol and ethylenediamine, hardly reduced the suspension viscosity. The PPG with Mw of 1,450, which is a branch in a comb-graft copolymer, also did not reduce the viscosity. The backbone in the comb-graft copolymer played an important role as a dispersant, producing a highly fluid SiC ethanol suspension. The SiC suspension with the lowest viscosity 40 mPa s was obtained using the copolymer from the AQNY backbone.

Table III. The apparent viscosity of a 20 vol.% SiC ethanol suspension made using various comb-graft copolymers with different types of backbones

Comb-graft Copolymer		Apparent Viscosity (mPa s)
Backbone	Mw of a PPG branch	
PEI-012	700	410
PEI-100	580	750
AQNY [1]	930	40
HPC [1]	630	4,000
Sorbitol [1]	730	3,050
Ethylenediamine [1]	690	3,200
Blank	---	5,000
PPG	1,450	4,930
PEI-012 [2]	---	35
PEI-100 [2]	---	14

1) Highly fluid suspensions were not obtained when using the original backbone compounds.
2) The original PEI as a backbone in the PEI comb-graft copolymer.

Table IV. The apparent viscosity of a 20 vol.% SiC ethanol suspension as a function of the molecular weight of a branch using various comb-graft copolymers

Comb-graft Copolymer		Apparent Viscosity (mPa s)
Backbone	Mw of a PPG branch	
PEI-012	--- (Original)	35
	60	230
	350	240
	720	410
	1,450	515
	2,880	655
AQNY	----	--- [1]
	930	40
	1,660	28
	3,200	1,250
HPC	----	--- [1]
	630	4,000
	1,440	60
	3,800	3,500
Blank		5,000

1) The suspension with fluidity was not obtained.

Effects of Branched Chains in a Comb-graft Copolymer

Because the copolymers from PEI and AQNY affected the suspension viscosity, the effects of the branched chains in a comb-graft copolymer were investigated. Copolymers containing PPG branches with different Mws ranging from 60 to 3,800 were examined. Table IV shows the apparent viscosity of a 20 vol.% SiC ethanol suspension with each copolymer fixed at 0.71 mg/m². In the case of the PEI copolymers, the suspension viscosity gradually increased with an increase in Mw from 60 to 2,880 (i.e., branch chain length). The AQNY copolymer containing PPGs with a Mw of 930 and 1,660 showed the lowest viscosities (i.e., 40 and 28mPa s), although the viscosity of the suspension of the AQNY copolymer with PPGs with a Mw of 3,200 increased up to 1,250 mPa s. The optimum branched chain length to achieve the minimum viscosity with the HPC copolymer was determined to be a Mw of 1,440. It seems that the HPC copolymer with short chains of PPG does not dissolve in ethanol as well as the original HPC. The length of branched chains in the comb-graft copolymer was related to a decrease in the viscosity of the SiC ethanol suspension, although the optimum length depended on each backbone polymer.

Comparisons of Comb-graft Copolymers with Original PEI

To determine the difference between the comb-graft copolymers and the original PEI, and to determine the effect of the backbones in the copolymers, the effect of the amount of dispersant in the ethanol suspension was investigated using the PEI, HPC and AQNY copolymers with PPGs of Mw of about 1,500 (which have good dispersibility). Figure 2 shows the viscosity of 20 vol. % SiC suspensions with the copolymers and the original PEI with a Mw of 5,000.

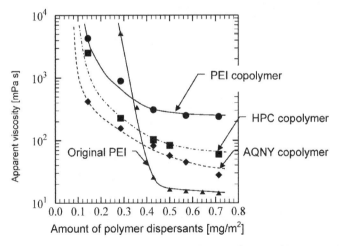

Figure 2. The apparent viscosity of 20 vol. % SiC ethanol suspensions as a function of the amount of the comb-graft copolymer

Compared to PEI, suspension viscosity gradually decreases with increasing amount of comb-graft copolymer from 0.14 to 0.71 mg/m². Fluid SiC suspensions were produced at about 0.23 mg/m², although the viscosity of the suspensions with the original PEI drastically decreased down to 15 mPa s at 0.42 mg/m². The suspension containing 0.71 mg/m² of the AQNY copolymer displayed the lowest viscosity at 28 mPa s. The effect of the copolymers on suspension viscosity differs depending on their backbone.

Based on these results, it is difficult to control the aggregation and dispersion behavior of particles in suspensions with the original PEI, because the suspension viscosity drastically changes with the additive content. However, the comb-graft copolymers are able to reduce the suspension viscosity with a low amount of polymer, and it is easy to control the suspension viscosity. As reported in our previous articles[19,20], the microstructure and strength of the spray-dried granules of the silicon nitride powders were shown to be a function of the state of the powder agglomeration in the suspension. When close-packed granules are found in finely dispersed suspensions, the granule strength is greater, and inter-granular pores remain in the green compacts and the final sintered bodies, which decreases the fracture strength of the sintered bodies.

The effect of suspension concentration on suspension viscosity with each copolymer at 0.71 mg/m² was investigated. Figure 3 shows the viscosity of the SiC ethanol suspensions with different solid volume fractions. The trends are consistent with Figure 2. With the original PEI, the suspension viscosity remains below 100 mPa s at almost a 46 vol. % solids. With the AQNY and HPC copolymers, the solids concentration must be less than 34 and 27 vol. %, respectively, to maintain a suspension viscosity under 100 mPa s. This low suspension viscosity cannot be achieved with the PEI copolymer. The effect of the suspension concentration depends on the backbone of the comb-graft copolymer. With the comb-graft copolymers, low viscosity suspensions can be produced with less polymer. However, relative to the original PEI, the copolymers are less effective in decreasing suspension viscosity.

The steric repulsion between particles with adsorbed polymer on particle surfaces promotes dispersibility in a nonaqueous suspension. Accordingly, the relationship between suspension viscosity and the amount of copolymer adsorbed onto a SiC particle surface was investigated. Table V shows the amount of adsorbed copolymer on the SiC particle in ethanol suspensions when using 0.71 mg/m² of copolymer. Low viscosity is observed when there is a relatively large amount of copolymer adsorbed onto the SiC particles. Relative to the original PEI, the amount of PEI copolymer adsorption decreases from 0.56 mg/m² down to 0.184 mg/m² as the Mw of the PPG branches increases. The amount of HPC and AQNY copolymer adsorption was greater than that of the PEI copolymer, and was 0.282 and 0.253 mg/m², respectively. In contrast, copolymers made from backbones with low Mws (i.e., sorbitol and ethylenediamine), and the branch, (i.e., PPG), hardly adsorb onto the SiC particles at all. These compounds do not act to reduce the SiC ethanol suspension viscosity. The backbone in a comb-graft copolymer significantly affects the amount of adsorption of the copolymer on a SiC particle. The adsorbed polymer will generally produce a low viscosity SiC ethanol suspension.

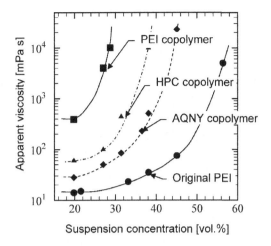

Figure 3. The apparent viscosity of SiC ethanol suspensions with different solid volume fractions

DISCUSSION

In general, the minimum suspension viscosity is achieved when the adsorption sites on a particle surface become saturated. Based on the results in Figure 2 and Table V, the minimum viscosity with the original PEI corresponds to the saturated adsorption concentration. However, viscosity decreases with increasing copolymer content over the adsorption saturation content (Table). It seems that the amount of adsorption increases with increasing copolymer content. Figure 4 shows the adsorption isotherms. With the original PEI, all of the polymer adsorbs onto the SiC particle surface up to the saturation concentration, after which the adsorption amount plateaus. This point corresponds to that of the minimum suspension viscosity. The copolymers also show a saturation concentration, although the absolute amount is lower than that with PEI. The saturation concentration depends on the backbone in the copolymer. However, the saturation concentrations cannot explain the observed viscosity behavior. Dispersability may also be affected by the free polymer, which is not adsorbed onto the particles.

Some functional groups in the polymer backbone, as is shown in Table I, can act as multiple adsorption sites on a SiC particle surface. To investigate the interaction between the backbone and a characteristic particle surface, the effect of the backbone in the comb-graft copolymer on the viscosity of the ethanol suspension with different ceramic powders (i.e., Al_2O_3, Si_3N_4, and SiC), was examined. These powders have different isoelectic points as shown in

Table II. Table IV shows the apparent viscosity of the ceramic ethanol suspensions made from copolymers with PEI, HPC and AQNY backbones. All suspensions with the HPC copolymer displayed high fluidity. The PEI copolymer is adsorbed only on the SiC particles, and reduces only the SiC suspension viscosity. The fluidity of the suspensions with each ceramic powder depended on the backbone in the copolymer. The polymer with the functional groups that strongly interact with active sites[21] on the particle surface seem to be the most effective backbone in the comb-graft copolymer as a dispersant.

Table V. The amount of adsorbed copolymer on a SiC particle in 20 vol.% ethanol suspensions with various comb-graft copolymers and different backbones, and with PPG branches with different molecular weights.

Comb-graft Copolymer		Amount of adsorbed copolymer (mg/m^2)	Apparent Viscosity (mPa s)
Backbone	Mw of a PPG branch		
PEI-012	720	0.229	410
PEI-012	1,450	0.194	515
PEI-012	2,880	0.184	655
AQNY	1,660	0.282	28
HPC	1,440	0.253	60
Sorbitol	1,620	0.004	2,700
Ethylenediamine	690	0.005	3,200
PEI-012 [1]	---	0.560	35
PPG	1,480	0.003	5,000

1) The original PEI as a backbone in the PEI comb-graft copolymer.

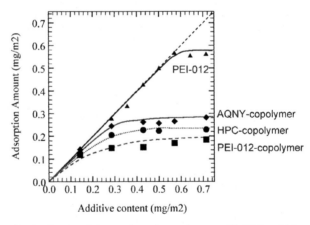

Figure 4. The adsorption isotherms of the comb-graft copolymers with PPGs of Mw of about 1,500 from AQNY, HPC and PEI, and the original PEI with Mw of 1,200

Table VI. The apparent viscosity of the 26% vol. Al_2O_3 and Si_3N_4 ethanol suspensions, and the 20% vol. SiC ethanol suspensions using the comb-graft copolymers, and the amount of PEI copolymer adsorption .

Comb-graft Copolymer		Apparent Viscosity (mPa s)		
Backbone	Mw of a PPG branch	Al_2O_3	Si_3N_4	SiC
PEI-012	1,450	4,050	>10,000	515
AQNY	1,660	8,500	390	28
HPC	1,450	510	380	60
Blank	---	7,100	10,300	5,000
Amount of PEI-12 copolymer adsorption (mg/m2)		0.0029	0.003	0.194

On the other hand, the branched chain PPGs of a comb-graft copolymer are not adsorbed onto the SiC particle surface, and will extend into ethanol. The PPGs will produce a steric repulsive force between the particles. However, the PPGs with excessive Mw prevent the polymer backbone from coming into contact with and adsorbing on a particle surface. Therefore, an increase in Mw of PPG will decrease PEI copolymer adsorption, and will reduce the dispersibility of the SiC particles. In general, the lowest viscosity suspension will be produced, when the dispersant concentration has reached the adsorption saturation concentration. In the case where the viscosity continues to decrease over the saturation concentration, it is believed that the free polymer affects the particle interactions and reduces the viscosity.

CONCLUSION

Using a series of comb-graft copolymers from five types of backbone compounds and polypropylene glycol (PPG) branches with different Mws, the effect of the backbone and the branched chains in a copolymer on the viscosity of SiC/ethanol suspensions was investigated. The copolymers from the polymer, such as polyethyleneimine (PEI), hydroxypropyl cellulose (HPC), and aqueous nylon (AQNY), adsorbed onto the SiC particles and reduced the suspension viscosity at low concentrations compared to the original PEI. It was easy to control the rheological properties of the suspension. Moreover, the HPC copolymer effectively dispersed Al_2O_3 and Si_3N_4 particles in ethanol suspensions. The copolymer from the backbone with a low Mw (i.e., sorbitol and ethylenediamine) did not absorb onto SiC, and did not reduce the suspension viscosity. The suspension viscosity depends on the backbone that adsorbs onto the particle, and the PPG branches that will produce steric repulsion to decrease the suspension viscosity. Each backbone in a copolymer had the optimum Mw (length) of PPG branches to achieve the minimum suspension viscosity. Based on adsorption isotherms and suspension viscosity, free polymer (i.e., copolymer not adsorbed onto particle surfaces) also decreases the suspension viscosity.

REFERENCES

[1]B. Siffert, A. Jada, and J. Eleli-Letsango, "Stability Calculations of TiO_2 Non-aqueous Suspensions: Thickness of the Electrical Double Layer," *J. of Colloid and Int. Science*, **167**, 281-286 (1994).

[2]B.Siffert, A. Jada, and A. Eleli-Letsango, "Location of the Shear Plane in the Electric Double Layer in an Organic Medium," *J. of Colloid and Int. Science*, **167**, 327-333 (1994).

[3]B. I. Lee, B. I., and U. Paik, "Dispersion of Alumina and Silica Powders in Non-aqueous Media: Mixed – Solvent Effects," *Ceramics International*, **19**, 241-250 (1993).

[4]J. H. H. ter Maat, "Critical Factors Determining the Activity of Oligomeric Dispersants," Ceramic Processing Science and Technology, *Ceramic Transactions*, 355-359 (1995).

[5]W.M. Sigmund, G. Wegner, and F. Aldinger, "Interaction of Organic Additives with Alumina Surfaces in a Ceramic Slurry," *Mat. Res. Soc. Symp. Proc., Materials Research Society*, **407**, 313-318 (1996).

[6]H. S. Al-Lami, N. C. Billingham, and P. D. Calvert, "Controlled Structure Methacrylic Copolymers as Dispersants for Ceramics Processing," *Chem. Mater.*, **4**, 1200-1207 (1992).

[7]L. Bergstom, "Rheological Properties of Concentrated Nonaqueous Silicon Nitride Suspensions," *J. Am. Ceram. Soc.*, **79**, 3033-3040 (1996).

[8]H. Kamiya, Y. Fukuda, Y. Suzuki, M. Tsukada, T. Kakui, T. and M. Naito, "Effect of Polymer Dispersant Structure on Electosteric Interaction and Dense Alumina Suspension Behavior," *J. Am. Ceram. Soc.*, **82**, 3407-3412 (1999).

[9]M. Nojiri, H. Hasegawa, T. Ono, T. Kakui, M. Tsukada, and H. Kamiya, "Influence of Molecular Structure of Anionic Polymer Dispersants on Dense Silicone Carbide Suspension Behavior and Microstructures of Green Bodies Prepared by Slip Casting," *J. Ceram. Soc. Japan*, **111**, 327-332 (2003).

[10]H. Kamiya, S. Matyui, and T. Kakui, "Analysis of Action Mechanism of Anionic Polymer Dispersant with Different Molecular Structure in Dense Silicon Nitride Suspension by using Colloidal Probe AFM," *Ceramic Transaction*, **152**, 83-92 (2003).

[11]T. Kakui, and H. Kamiya, "Effect of Polymer Dispersant Molecular Structure on Ceramics Suspension Behavior in Nonaqueous System," *Ceramic Transaction*, **146**, 43-50(2004).

[12]T. Kakui, T. Miyauchi, and H. Kamiya, "Analysis of the Action Mechanism of Polymer Dispersant on Dense Alumina Suspension Behavior in Non-aqueous System using Colloidal Probe AFM," *J. Euro. Ceram. Soc.*, **25**, 655-661 (2005).

[13]G. H. Kirby, and J. A. Lewis, "Comb Polymer Architecture Effects on the Rheological Property Evolution of Concentrated Cement Suspensions," *J. Am. Ceram. Soc.*, **87**, 1664-1652 (2004).

[14]N. Tobori, and T. Amari, "Rheological Behavior of Highly Concentrated Aqueous Calcium Carbonate Suspensions in the Presence of Polyelectrolytes," *Colloids and Surfaces A: Physicochem. Eng. Aspects*, **215**, 163-171 (2003).

[15]H. Yoshihara, "Graft Copolymers as Dispersants for CWM, "*Coal Preparation (Gordon & Breach)*, **21**, 93-103 (1999).

[16]F. Bonaccorsi, A. Lezzi, A. Prevedello, L. Lanzini, and A. Roggero, "Synthesis of Poly-(4-methylstyrene)-graft-poly (ethylene oxide) s and Application as Polymeric Dispersant in Coal Water Mixture," *Polymer International*, **30**, 93-100 (1993).

[17]T. Nakanishi, K. Furuya, A. Mishida, M. Hirao, H. Itoh, S. Tatsumi, H. Ozaki, S. Takano, and Y. Kajibata, "Evaluation of Dispersing Additives for Coal Water Mixture," *Proceedings of 14th Int. Cong. on Coal & Slurry Tech.*, 321-330 (1989).

[18]A. Naka, Y. Nishida, H. Sugiyama, and T. Sugiyama, "Ability of N-[poly(oxyethylene)] - polyethyleneimine to increase the Coal Content of Coal Water Slurry," *J. Chem. Soc. Jpn.*, **13**, 227-230 (1989).

[19]H. Kamiya, K. Isomura, G. Jimbo, and J. Tsubaki, "Powder Processing for the Fabrication of Si_3N_4 Ceramics: I, Influence of Spray-Dried Granule Strength on Pore Size Distribution in Green Compacts," *J. Am. Ceram. Soc.*, **78**, 46-57 (1995).

[20]H. Kamiya, M. Naito, T. Hotta, K. Isomura, J. Tsubaki, and K. Uematsu, "Powder Processing for the Fabrication of Si_3N_4 Ceramics: II," *Am. Ceram. Soc. Bull.*, **76**, 79 (1997).

[21]R. J. Pugh, "Surface Acidity/Basicity of SiC, ZrO_2, α-Al_2O_3 and Y_2O_3 Ceramic Powders," *Ceramic Powder Science (Ceramic Transactions)*, 375-382 (1990).

DISPERSION CONTROL AND MICROSTRUCTURE DESIGN OF NANOPARTICLES BY
USING MICROBIAL DERIVED SURFACTANT

Hidehiro Kamiya, Yuichi Iida, Kenjiro Gomi, Yuichi Yonemochi, Shigekazu Kobiyama,
Motoyuki Iijima and Mayumi Tsukada
Graduate School of Bio-Applications and Systems Engineering
Tokyo University of Agriculture & Technology
Koganei, Tokyo 184-8588, Japan

ABSTRACT

To produce uniformly dispersed $BaTiO_3$ and $CaCO_3$ particles in the size range of several
tens of nanometers, a special, new microbial derived aqueous surfactant with a high hydrophilic
group and a cis-structure, was added to the reactant solution before particle synthesis. By adding
the surfactant, it is possible to decrease the size of aggregates in suspension, and improve the
dispersion stability of $BaTiO_3$ and $CaCO_3$ nanoparticle suspensions. The role of this surfactant
on particle formation was determined by FT-IR analysis completed on a mixture of $Ba(OH)_2$
aqueous solution and the dispersant. It was confirmed that carboxyl (COO^-) groups in the
surfactant form a complex with the Ba2+ ions. Based on this result, it was estimated that pools
on the order of several tens of nanometers are formed among the complexes of the surfactants
and the metal ions in a high pH solution. This is attributed to the bending structure of the
microbial derived surfactant and the strong electrostatic repulsive force between surfactants. In
the synthesis of $CaCO_3$ nanoparticles, it was also possible to control the crystal phase with the
addition of the microbial derived surfactant.

INTRODUCTION

Many researchers are currently investigating new processes to synthesize nanometer size
ultrafine inorganic particles from high-purity raw materials (e.g., sol-gel and the hydrothermal
processes). Using these processes, it is rather easy to prepare stoichiometric composition, single
phase, and less than 100 nm in diameter barium titanate[1], calcium carbonate[2], and other
inorganic functional particles. However, there is rapid sedimentation in suspension because of
the rapid formation of large, porous aggregates that are several tens of micrometers in diameter.

In our previous research[3,5], we succeeded to prepare barium titanate nanoparticles with
high dispersion stability in suspension by the addition of a microbial-derived surfactant[4] and an
artificial surfactant[5] during the wet process of particle synthesis. The addition of each surfactant
changed the packing microstructure of the aggregates from a branch-shape structure to closed
packing structure of ultrafine particles, and it was possible to decrease the size of the aggregates
in suspension and to obtain stable dispersion of $BaTiO_3$ nanoparticles in aqueous suspension.

In this paper, FT-IR was used to analyze the role of this microbial derived surfactant in a
$Ba(OH)_2$ aqueous solution before the synthesis of $BaTiO_3$ nanoparticles. Additionally, the effects
of adding this surfactant to an aqueous suspension of $BaTiO_3$ nanoparticle prepared from a
reactant solution without this surfactant was studied. The sedimentation behavior of both
suspensions was compared. Based on the efficiency of the surfactant to produce stable
suspensions, $CaCO_3$ nanoparticles were then synthesized by precipitation with or without the
addition of this surfactant.

EXPERIMENTAL
Preparation and Characterization of BaTiO₃ Powder

Barium hydroxide octahydrate and titanium tetraisopropoxide were used as starting materials for ultra-fine $BaTiO_3$ powder synthesis. The purities of the barium hydroxide octahydrate and titanium tetraisopropoxide was 98% and 95%, respectively, and they were provided by Wako Pure Chemical Industries, Japan. A 0.4M isopropyl-alcohol solution of titanium tetraisopropoxide was prepared by diluting 11.4g of titanium tetraisopropoxide in isopropyl alcohol (Wako Pure Chemicals Industries, Japan) to 100ml. A 0.2M barium hydroxide aqueous solution was prepared by mixing 10.1g of barium hydroxide octahydrate, 156.7ml of distilled water, and 10.7ml of 12N sodium hydroxide (Wako Pure Chemicals Industries, Japan) solution. The pH of the suspension that was made by the reaction of the barium hydroxide aqueous solution with the sodium hydroxide and the isopropyl alcohol solution of titanium tetraisopropoxide was about 13. Since the isoelectric point of $BaTiO_3$ is about 6 or 7, the ultra-fine powder produced in the suspension was negatively charged.

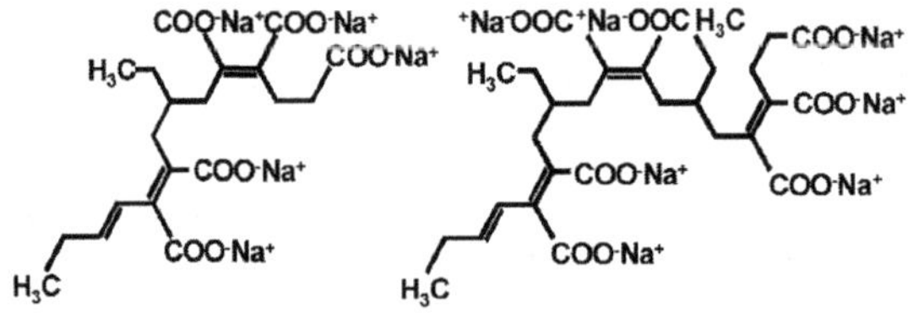

(a)   (b)

Figure 1. Molecular structure of microbial derived surfactant
(Surfactant is the mixture of (a) and (b) with the molar ratio of a:b = 1:1).

A microbial derived surfactant, whose molecular structure is shown in Figure 1, was used as a surfactant. Using the surfactant in the same preparation method as in our previous work[3], $BaTiO_3$ nanoparticles were prepared. The surfactant was either added to the 0.2M barium hydroxide aqueous solution before precipitation (denoted as "Method 1" in this paper), or added to the aqueous suspension after the synthesis of the $BaTiO_3$ nanoparticles in reactant solutions without the surfactant. The process of first forming the nanoparticles and subsequently adding the surfactant, is denoted as "Method 2". In this paper, the concentration of the surfactant was fixed at 7.1 wt% relative to stoichiometric $BaTiO_3$. This is the optimum concentration for high dispersibility, and for keeping the $BaTiO_3$ mono-crystalline.

To examine the role of the surfactant on particle growth and dispersibility, the sediment, i.e. the settled mixture of the surfactant with the barium hydroxide aqueous solution) was analyzed by FT-IR. To determine the amount of adsorbed surfactant on the synthesized particles, the supernatant above the sediment was dried at 120°C for 12h to evaporate ethanol and water, and the remaining solid was dissolved with de-ionized water. Since the thermal decomposition temperature of the surfactant is about 450°C, it was assumed that the aqueous solution contained only surfactant and sodium hydroxide. The aqueous solution was analyzed by a HPLC, and the amount of surfactant adsorbed on synthesized particles was determined.

The sedimentation behavior of the suspensions prepared using Method 1, Method 2, and without surfactant was compared as follows. 100ml of synthesized suspension was transferred into a graduated cylinder where it stood in a vibration-proof state, and the time-dependent change in the volume of the sediment was measured.

Preparation and Characterization of $CaCO_3$ Powder

A batch type crystallizer was used to prepare the calcium carbonate particles. Four square Teflon baffle-plates were mounted inside the wall of the stirred reactor. In a separable flask in a water bath at 298 K, the reaction solution was strongly agitated with a six-blade impeller. Calcium chloride dihydrate ($CaCl_2 \cdot 2H_2O$), sodium carbonate dihydrate ($Na_2CO_3 \cdot 2H_2O$), and sodium hydroxide were used as starting materials for the preparation of the fine calcium carbonate powder. Each raw material was obtained from Wako Pure Chemical Industries, Japan. The purities of calcium chloride, sodium carbonate, and sodium hydroxide were >97 %, >99 % and >97 %, respectively.

Hirasawa et al.[2] used the batch precipitation process to form calcite by directly mixing calcium chloride and sodium carbonate solutions at a pH of about 13. In this paper, we used the same preparation conditions as Hirasawa et al.. 1.0 M aqueous solution of calcium chloride was prepared by dissolving 1.66 g of calcium chloride dihydrate into 15 ml of distilled water (Milli-Q, Millipore Corporation). 0.05 M sodium carbonate aqueous solution with a high pH was prepared by mixing 1.25 g of sodium carbonate dihydrate with 300 ml of distilled water and 60 ml of 2.0 M sodium hydroxide aqueous solution.

The microbial derived surfactant was added to the mixture of 0.05 M sodium carbonate and 2.0 M sodium hydroxide aqueous solution. In the present study, the additive content of the surfactant was indicated by the molar ratio of COO^- / Ca^{2+}. To determine the optimum surfactant content to prepare fine powder while maintaining high dispersion stability and homogeneous crystallization, the ratio of surfactant to calcium carbonate particles, i.e., the ratio of COO^- / Ca^{2+}, was changed within the range from 0 to 1.28.

To characterize the primary particle structure, after aging, the synthesized suspension was filtered and washed with de-ionized water a few times to remove the sodium ions in the filtered cake. This filtered cake was dried at 100℃ for 12h. The morphology and crystalline phases and the chemical composition of the dried powder sample were characterized by FE SEM and X-ray diffraction (RAD-Ⅱ C, Rigaku, Tokyo, Japan), respectively.

RESULTS AND DISCUSSION

Analysis of the Role of Surfactant in $BaTiO_3$ Preparation

In order to discuss the dispersion stability and aggregation behavior of synthesized $BaTiO_3$ ultra-fine particles in suspension, the sedimentation behavior of the suspensions prepared using the different conditions was measured and is shown in Figure 2. For the suspension without the addition of a surfactant, the sedimentation of the particles in suspension started immediately after the solution was transferred to the graduated cylinder, and the sedimentation of the ultra-fine $BaTiO_3$ particles was almost complete within 40min.

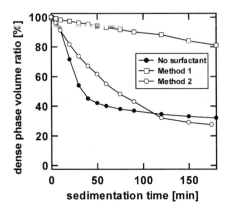

Figure 2. Sedimentation behavior of synthesized BaTiO$_3$ particles in suspension with without surfactant, and with surfactant using Method 1 and Method 2.

For Method 1, in which the surfactant was added before the synthesis of the nanoparticles, the rate of sedimentation of the particles in suspension was reduced. For Method 2, where the surfactant was added after the synthesis of BaTiO$_3$ ultra-fine particles, the sedimentation rate of particles in suspension was reduced, but the final volume of the sediment is almost same as that without the surfactant.

In our previous work[3] BaTiO$_3$ particle suspensions prepared using Method 1 and without surfactant were freeze-dried and observed by FE SEM. It was concluded that the addition of surfactant by Method 1 reduced the size of the aggregates from several 10s of micrometers to 100 ~ 200nm in diameter, and changed the packing structure of the primary particles in the aggregates from sparse and branch-shaped to the tightest possible one. The reduction in the size of the aggregates decreased the sedimentation rate of the aggregates in suspension, as shown in Figure 2.

To analyze the adsorption behavior of the surfactant, the amount of surfactant absorbed on the particles was measured and is shown in Table 1. Since almost all of the surfactant (93.5 %) was adsorbed on the BaTiO$_3$ particles prepared using Method 1, a high suspension stability was observed. However, for Method 2, the amount of adsorbed surfactant was reduced up to 20 %, even after extending the mixing time by up to 72 hours. It seems that the longer mixing time after the surfactant addition promoted the desorption of the surfactant.

Table 1. Effect of surfactant addition condition on the amount of adsorbed surfactant on the BaTiO$_3$ particles (Additive content of surfactant is 7.1 wt%).

Addition condition	Method 1	Method 2 (as a function of mixing time)			
		1 hr	3 hr	24 hr	72 hr
Surfactant Adsorbed	93.5 %	26.4 %	23.6 %	23.8 %	18.5 %

To produce a stable dispersion of nanoparticles, the surfactant must be added to Ba(OH)$_2$ aqueous solution. Since the sediment forms rapidly after the addition of the surfactant to the Ba(OH)$_2$ aqueous solution and before the reaction with the barium hydroxide octahydrate and titanium tetraisopropoxide, the sediment was dried and analyzed by FT-IR. The FT-IR spectra of the dried microbial derived surfactant alone and of the surfactant mixed with the Ba(OH)$_2$ aqueous solution (b) are shown in Figure 3. The wave number of absorbance of COO-Na at 1565 cm^{-1} was shifts to 1542 cm^{-1} in sediment indicating that the carboxyl (COO$^-$) groups in surfactant form a complex with the Ba^{2+} ion.

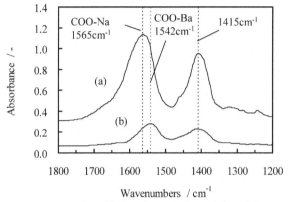

Figure 3. FT-IR analysis of (a) dried surfactant before mixing with Ba(OH)$_2$ aqueous solution, and (b) dried sediment produced after mixing surfactant with Ba(OH)$_2$ aqueous solution.

Based on this result, it was postulated that nanometer scale pools form among the complexes with the surfactant and positive ions in a high pH solution (Figure 4) because of the bending structure of the microbial derived surfactant and the strong electrostatic repulsive force between surfactants. Since the nucleation and growth of particles occurs in these nanometer scale pools, particle dispersion is promoted, and the surfactant adsorbs on the particles.

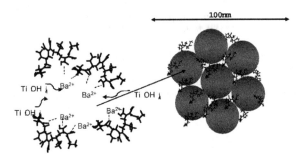

Figure 4. Illustration of the nanometer scale pools that are believed to form nanometer scale pools form among the complexes with the surfactant and positive ions in a high pH solution.

Effect of Surfactant Addition on Sedimentation Behavior of $CaCO_3$ Nanoparticles

If the surfactant forms a complex with a bivalent or trivalent positive ions in solution during the nucleation process, it can be expected that that highly dispersed nanoscaled particles can be synthesized from alkaline metal ions such as Mg^{2+}, Ca^{2+} or other metal ions such as Fe^{2+}, Fe^{3+}, Cu^{2+}, Pb^{2+} etc. In order to assess the dispersion stability and aggregation behavior of fine calcium carbonate particles in suspension, the sedimentation behavior of suspension prepared with different amounts of microbial derived surfactant was measured, and shown in Figure 5. The rapid sedimentation of the particles in the suspension without the surfactant was observed, and a stable dense layer formed within 30 min. The addition of the surfactant significantly reduced the rate of particle sedimentation in the suspension, and produced a stable colloidal suspension. The sedimentation rate decreased with increasing amount of surfactant.

The addition of the surfactant in the preparation of the calcium carbonate, controlled the crystal phase and morphology of primary calcium carbonate. Figure 6 shows the XRD patterns for the ultra-fine $CaCO_3$ powders synthesized, and the effect of the surfactant content on the crystal phase. For a COO^-/ Ca^{2+} molar ratio of 0.64, only calcite type calcium carbonate is observed in product powder, and the surfactant did not inhibit calcite synthesis. On the contrary, at a relatively high additive content higher than the molar ratio of 1.28, the peaks of calcite phase disappeared, and the vaterite phase is observed.

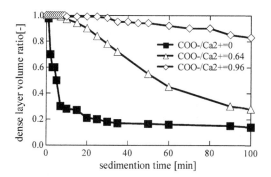

Figure 5. Sedimentation behavior of calcium carbonate without and with the addition of surfactant.

The vaterite phase is metastable in nature, indicating that the microbial derived surfactant regulates crystal phase formation. Previously in this paper, it was confirmed that the surfactant forms a complex with Ba^{2+} ions in a highly alkali solution. Based on this result, we expect the surfactant also forms a complex with the calcium ion. It seems that the nucleation and growth of $CaCO_3$ starts from this complex. Since the surfactant has a bending cis-structure, the surfactant is able to maintain the distance between nuclei, and behave as a rigid template to inhibit calcite phase formation and to selectively form the vaterite phase.

To examine the influence of the surfactant on the morphology of the primary calcium carbonate particles, FE SEM images of the product powders synthesized in solution with different amounts of surfactant are shown in Figure 7. The primary particle diameter observed by FE SEM is about 50 nm without surfactant, as shown in Figure 7 (a). When surfactant was added up to 0.96, vaterite is the only crystalline phase observed. A clear difference in primary particle morphology and anisotropic growth aspect, can be seen by FE SEM as shown in Figure 7 (b).

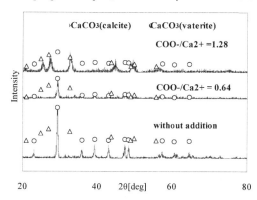

Figure 6. XRD patterns of ultra-fine $CaCO_3$ powders synthesized with and without surfactant.

 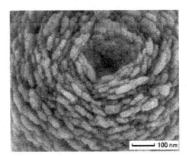

(a) without addition of surfactant   (b) with surfactant
$$(COO^- / Ca^{2+} = 0.96)$$
Figure 7. FE SEM micrographs of calcium carbonate particles synthesized without and with surfactant.

CONCLUSION

The mechanism for producing a stable dispersion of ultrafine $BaTiO_3$ particles using a microbial derived surfactant was discussed on the basis of various kinds of measurement and analyses. The surfactant promotes the formation of a complex with a bivalent or trivalent positive ion and carboxyl groups (COO^-) in aqueous solution before particle synthesis, and nanometer scale pools are produced amongst the networks of complexes. It is believed that the surfactant controls the nucleation and growth of the particles produced in these nanometer scaled pools, and adsorbs onto the particles to enhance particle dispersion, In order to confirm this hypothesis, the surfactant was used in the preparation of calcium carbonate nanoparticles. The addition of the surfactant improved the suspension stability and changed the crystal phase and particle morphology.

ACKNOWLEDGEMENT

This study was supported by a Grant-in-Aid for Scientific Research (B) from the Japanese Ministry of Education, Science, Sports and Technology, and the Murata Science Promotion Foundation, and the Structurization of Material Technology Project in the Nano-Technology Program by METI Japan.

REFERENCES

[1] K. W. Kirby, "Alkoxide Synthesis Techniques for $BaTiO_3$," *Materials Research Bulletin*, 23, 881-90 (1988).

[2] I. Hirasawa, A. Okazaki, S. Kurakata and K. Shima, "nm Size Control of Calcite from Amorphous State Triggered by Heterogeneous Substance", *Chemical Engineering Transactions*, 1, 179-184 (2002)

[3] H. Kamiya, K. Gomi, Y. Iida, K. Tanaka, T. Yoshiyasu and T. Kakiuchi, "Preparation of Highly Dispersed Ultra-Fine Barium Titanate Powder by Using Microbial-Derived Surfactant," *J. Am. Ceram. Soc.*, 86, 2011-2018 (2003).

[4] R. Yamanishi, K. Okada, N. Tamugi, M. Iwashima, and K. Iguchi, "A Novel Maleic Anhydride

Derivative from the Fungus Talaromyces sp. No. 10092," *Bull. Chem. Soc. Jpn.*, **73**, 2087-91 (2000).

[5]Y.Yonemochi, Y. Iida, K.Ogino, H. Kamiya, K. Gomi and K. Tanaka, "Preparation of Highly Dispersed Ultra-fine Barium Titanate powder by Using Acrylic Oligomer with High Density of Hydrophilic Group, *Ceramic Transactions*, **152**, 27-35 (2004)

Forming

ANALYSIS OF CONSOLIDATION BEHAVIOR OF 68 nm YTTRIA-STABILIZED ZIRCONIA PARTICLES DURING PRESSURE FILTRATION

Yoshihiro Hirata and Yosuke Tanaka
Department of Advanced Nanostructured Materials Science and Technology
Graduate School of Science and Engineering, Kagoshima University
1-21-40 Korimoto, Kagoshima 890-0065, Japan

ABSTRACT

The applied pressure and suspension height during the consolidation of an aqueous suspension containing 68 nm yttria-stabilized zirconia (YSZ, isoelectric point pH 2.8) particles at pH 3.0-9.2 in a closed cylinder were continuously recorded using a pressure filtration apparatus. The apparent viscosity became higher at pH 6-9. The suspension was filtered through three sheets of a 0.1 µm pore diameter membrane filter attached to the bottom of the piston displaced at a constant rate of 200 µm/min. The solution flowed into the pore channels in the upper piston in the Teflon cylinder. The final packing density of the 22 vol% suspension at pH 3.0-6.1 reached 50-53 % at 19 MPa of applied pressure. The decrease of the solid content to 5 vol% in the viscous suspension at pH 9.2 was effective to increase the packing density. The energy required to consolidate a unit volume of YSZ particles increased gradually with increasing volume of the dehydrated solution. The consolidation energy for 1 cm^3 YSZ particles was in the range from 2.43 to 8.18 J, and became larger for the suspension at pH 6.1. Because of the release of the stored elastic strain energy, the height of the consolidated YSZ compact increased after releasing the pressure. This resulted in a decrease in the packing density (47-48 %).

INTRODUCTION

It has been well recognized that colloidal processing, which is composed of the dispersion of a starting powder in a liquid media and subsequent consolidation, is superior to conventional dry pressing in the control of density and microstructure of green and sintered compacts.[1-7] The consolidation rate of colloidal particles and the structure of a consolidated powder cake are affected by particle size, the concentration of particles, the interaction energy between the colloidal particles, and the rheological properties. That is,

73

the forming is an important process as well as the dispersion of colloidal particles. However, few papers have reported on the consolidation energy (consolidation pressure) of colloidal particles.[8-10] It is reported that the packing density of a flocculated suspension is proportional to the logarithm of applied pressure in pressure filtration.[8-10] This result is related to the compressibility of the flocculated colloidal cake. A small effect of applied pressure on packing density has been reported for well dispersed .[8-11] In this paper, the consolidation behavior of a flocculated or dispersed yttria-stabilized zirconia (YSZ) suspension was measured in a newly developed pressure filtration apparatus. The suspension was filtered through three sheets of a 0.1 μm pore diameter membrane filter attached to the bottom of a piston in a closed cylinder. When the piston moved to compress the suspension, the filtrate flowed into and through the pore channels in the piston. The measured suspension height as a function of applied pressure was used to determine the relationship between the consolidation energy and suspension concentration.

EXPERIMENTAL PROCEDURE

A high purity YSZ powder supplied by Tosoh Co. Ltd., Japan, was used in this experiment: ZrO_2 86.57 mass%, Y_2O_3 13.42 mass%, SiO_2 0.003 mass%, Al_2O_3 0.005 mass%, Fe_2O_3 0.004 mass%, median size 67.8 nm, specific surface area 14.8 m^2/g. The zeta potential of the as-received YSZ powder was measured as a function of pH at the constant ionic strength of 0.01-M NH_4NO_3 (Rank Mark , Rank Brothers Ltd., UK). The colloidal suspensions of 22 vol% YSZ were prepared at pH 3.0, 5.1 and 6.1. Since the viscosity of the suspension at pH 9.2 was high, more dilute suspensions of 5, 10 and 15 vol% YSZ were prepared. The pH of the suspension was adjusted using 0.1 M-HCl or 0.1 M-NH_4OH solution. The rheological behavior of the suspensions was measured with a cone- and plate-type viscometer (Model EHD type, Tokimec, Inc., Tokyo, Japan).

Figure 1 shows the schematic illustration of the pressure filtration apparatus. The suspension in a closed cylinder was filtered through a plastic filter with 20 μm pore diameter and three sheets of a membrane filter with 0.1 μm pore diameter in a pressure range from 0 to 19 MPa. These filters were attached to the bottom of the piston (polymeric resin) moving at a crosshead speed of 200 μm/min to compress the suspension. The filtrate flowed into and through the pore channels in the upper piston. The applied load and the height of the piston were continuously recorded (Tensilon RTC, A & D Co., Ltd, Tokyo, Japan). The measured applied load – suspension height curve was integrated to obtain the energy for the consolidation of colloidal YSZ particles.[12] The

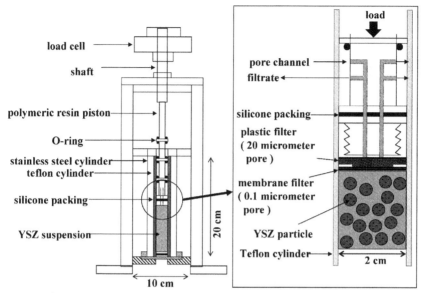

Fig.1   Schematic illustration of pressure filtration apparatus.

consolidated YSZ compact was taken out of the cylinder and dried at 100 °C in air for 24 h. The dried compact was heated at 600 °C in air for 4 h to give an enough strength for the measurement of bulk density by the Archimedes method using distilled water.

RESULTS AND DISCUSSION

Interaction energy between YSZ particles

The YSZ particles were charged positively at 4.9 mV at pH 2.1 and negatively above pH 3, and the isoelectric point was pH 2.8. In the pH range from 6 to 10, the change in zeta potential is small (-25.6 mV at pH 6.0 and -32.5 mV at pH 10.0). To understand the interaction between the YSZ particles, the van der Waals attraction energy (Ea) and the repulsion energy (Er) due to the electric double layer were calculated. Equation (1) represents the potential energy Ea,[6,13]

$$Ea = -\frac{A}{12}\left[\frac{D^2}{H^2 + 2DH} + \frac{D^2}{(H + D)^2} + 2\ln\frac{H^2 + 2DH}{(H + D)^2}\right] \tag{1}$$

where A and H are the Hamaker constant and the distance between two particles of diameter D (67.8 nm), respectively. As a Hamaker constant for ZrO_2, 7.23×10^{-20} J is reported.[14] Equation (2) corresponds to the repulsion energy between two charged particles,

$$Er = 32\pi\varepsilon\varepsilon_0 \left(\frac{D}{2}\right)\left(\frac{RT}{ZF}\right)^2 \tanh^2\left(\frac{ZF\varphi}{4RT}\right) \ln\left[\sqcap \exp\left(-\kappa H\right)\right] \qquad (2)$$

where ε is the relative dielectric constant (78.3 for H_2O), ε_0 is the permittivity of vacuum (8.854×10^{-12} F/m), R is the gas constant (8.314 J/mol K), T is the temperature (assumed to be 298 K), Z is the charge number of the electrolyte (assumed to be +1), F is the Faraday constant (9.649×10^4 C/mol), φ is the surface potential (approximated by the zeta potential) and $1/\kappa$ is the double layer thickness (assumed to be 10 nm).

Figure 2 shows the interaction energy, Ea + Er, for the YSZ particles as a function of the distance H between two YSZ particles. In the suspension at pH 3, two particles form a cluster by the attractive force when they approach within 10 nm. In the suspensions at pH 5-9, a repulsive interaction is expected within 20 nm. Figure 3 shows the apparent viscosity at 76.7 s^{-1} of shear rate for the YSZ suspensions at pH 3.0, 5.1, 6.1 and 9.2. The viscosity of the 22vol% YSZ was low at pH 3.0 and 5.1 and increased at pH 6.1. The higher viscosity was measured in the suspension at pH 9.2. The distance (H) between two particles was estimated from Eq.(3) using a random close packing model, [15,16]

$$H = D\left[\left(\frac{1}{3\pi C} + \frac{5}{6}\right)^{0.5} - 1\right] \qquad (3)$$

where C is the concentration (vol%) of the suspension. The H value becomes 0 at C = 63.7 vol%. The H value was calculated to be 49, 26, 16 and 10 nm for the solid content of 5, 10, 15 and 22 vol%, respectively. The interaction energy for the corresponding particle distance is shown in Fig.2. The low viscosity for the suspensions of 22 vol% YSZ at pH 3.0 and 5.1 is explained by the small interaction energy and the repulsive energy at H = 10 nm, respectively. However, it is difficult to explain the higher viscosity at pH 6 and 9 from the corresponding interaction energy shown in Fig.2. The possible explanation for the higher viscosity under the shear stress is the formation of particle agglomerates through dehydration and condensation between hydrated surfaces: $Zr(OH)_4 + Zr(OH)_4 \rightarrow Zr(OH)_3-O-Zr(OH)_3 + H_2O$. This reaction repeats three dimensionally. The remaining hydroxyl reacts with NH_4OH solution to form the

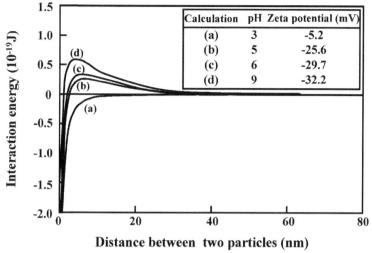

Fig.2　Calculated interaction energy for YSZ particles as a function of distance between two particles

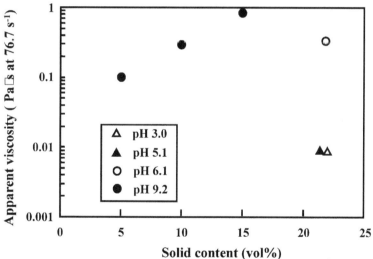

Fig.3　Apparent viscosity of YSZ suspensions at 76.7 s^{-1} of shear rate and solid content.

negatively charged ZrO^- sites ($-ZrOH + NH_4OH \rightarrow -ZrO^- + NH_4^+ + H_2O$). Figure 4 shows the infrared spectra of YSZ powders, which were consolidated from 10 vol% YSZ suspension and dried at 100 °C for 24 h. The infrared spectrum of the as-received YSZ powder was also measured as a reference datum. The stretching and bending

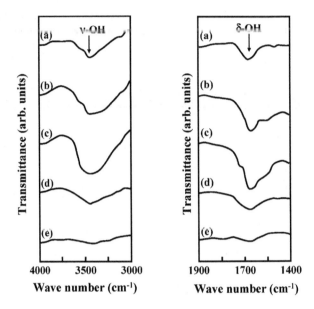

Fig. 4 Infrared spectra of (a) as-received YSZ powder, which was dried at 100 °C for 24 h, and YSZ powders dried from 10 vol% suspensions at (b) pH 3.0, (c) pH 5.1, (d) pH 6.1 and (e) pH 9.2.

vibration of hydroxyls were measured around 3450 and 1640 cm^{-1} for the YSZ in suspension at pH 3.0 and 5.1, and for the as-received YSZ powder. Additionally, the absorption peaks for the suspensions at pH 6.1 and 9.2 decreases. This result is associated with the dehydration and condensation between $Zr(OH)_4$ surfaces, and supports the formation of particle agglomerates.[17] In an acidic solution, the surface hydroxyl reacts with HCl solution to form the positively charged $-ZrOH_2^+$ site ($-ZrOH + HCl \rightarrow -ZrOH_2^+ + Cl^-$). The increased number of positively charged $-ZrOH_2^+$ sites shifts the zeta potential to the direction of positive value. Furthermore, the solubility of hydrated surface in an acidic solution is higher than that in a neutral or basic solution ($Zr(OH)_4 \rightarrow Zr^{4+} + 4OH^-$).[18] This tendency prevents the chemical interaction between the hydrated surfaces, leading to the increase in the dispersibility of YSZ particles.

Pressure filtration of YSZ suspension

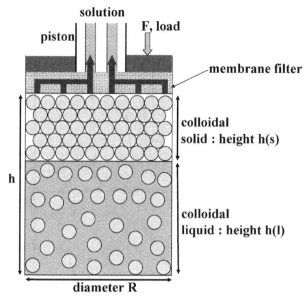

Fig.5   Schematic model of powder consolidation in the pressure filtration apparatus shown in Fig.1.

Figure 5 shows the schematic model for the filtration of the YSZ suspension. The dehydrated solution is drained toward the upper direction to form a particle cake below the membrane filter. The solid concentration of the suspension (the average concentration for the colloidal solid and colloidal liquid) and the volume of the dehydrated solution are given by Eqs. (4) and (5), respectively,

$$C = (\frac{h_0}{h}) C_0 \tag{4}$$

$$V = \pi (\frac{R}{2})^2 (h_0 - h) = \pi (\frac{R}{2})^2 h_0 (\frac{C - C_0}{C}) = V_0 (\frac{C - C_0}{C}) \tag{5}$$

where h_0 is the height of initial suspension, h is the height of suspension during the pressure filtration, C_0 is the initial concentration of suspension and R is the diameter of piston, and V_0 is the initial volume of the suspension in the cylinder. The height of the consolidate cake (h(s)) is expressed by Eq. (6),

$$\frac{h(s)}{h} = \frac{C - C_0}{C_f - C_0} = \frac{C_0}{C_f - C_0} (\frac{V}{V_0}) (\frac{1}{1 \; (\frac{V}{V_0})}) \tag{6}$$

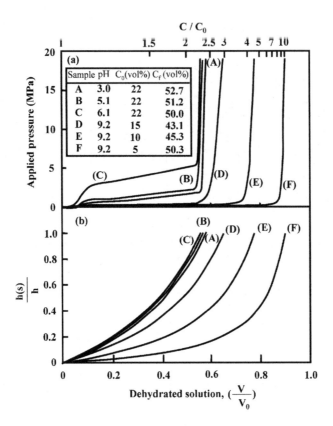

Fig. 6 Relationship between normalized volume of the dehydrated solution (V / V$_0$) and (a) applied pressure and (b) h(s) / h ratio. See Eqs.(5) and (6) in text for V / V$_0$, and h(s) / h ratios.

where C$_f$ is the packing density of YSZ particles at 19 MPa of applied pressure.

Figure 6 shows the relationship between the normalized volume of the dehydrated solution (V/V$_0$) and (a) applied pressure and (b) h(s) / h ratio. A very low pressure was measured to filtrate double distilled water, indicating a smooth flow of solution into the glass filter and membrane filter. For the consolidation of the 22 vol% YSZ concentration to V/V$_0$ = 0.54 (48 vol% solid), a low pressure of 2.5 MPa was measured for the suspensions at pH 3.0 and 5.1. The suspension at pH 6.1 needed a higher applied pressure of 5 MPa. Further increases in the concentration of the YSZ suspension to 50-53 % solid

(V / V_0 = 0.56-0.58) required a rapid increase in the pressure up to 19 MPa. The h(s) / h ratios increased nonlinearly with increasing V value. The reproducibility of the experimental result was very high. The rapid increase of the applied pressure at the final stage of the consolidation expresses the elastic properties of the consolidated cake under the compressive force. When the applied pressure was released, the height of the consolidated cake increased because of the release of the stored elastic strain energy. The degree of relaxation was 8.5-10% in the height of the compact. The packing density after the relaxation (47-48 % density) agreed with the bulk density of YSZ compact measured by the Archimedes method after calcination at 600 °C. Figure 6 also shows the consolidation behavior of the viscous YSZ suspension at pH 9.2. This suspension formed a cake with a low applied pressure before the final stage of the consolidation. The packing density of the suspension at 19 MPa was 50.3, 45.3 and 43.1% at 5, 10 and 15 vol% YSZ, respectively. These values of the packing density were lower than those for the suspensions at pH 3.0-6.1. The decrease in the solid content (increase of the distance between interacted particles) of the viscous suspension increases the packing density.

Consolidation energy

The area surrounded by the applied pressure − dehydrated solution curve (Fig. 6) corresponds to the energy for the consolidation of particles. The consolidation energy (Ec) for 1 cm^3 particles increased gradually with increasing V value. In the 22 vol% suspensions, the Ec values just before the elastic deformation of the consolidated cake at V / V_0 = 0.52-0.53 (packing density of 47-48 %) were 2.43, 3.33 and 8.18 J / cm^3-YSZ powder for the suspensions at pH 3.0, 5.1 and 6.1, respectively. The Ec of the suspension at pH 6.1 was relatively higher than that for the suspension at pH 3.0 or pH 5.1. The Ec values of the suspensions of 5-15 vol% YSZ at pH 9.2 were 1.61-3.98 J / cm^3-powder. The energy (W) applied between two particles during the consolidation was approximated by Eq. (7),[19]

$$W = \frac{2Ec}{NZ} \tag{7}$$

where E_c is the consolidation energy of the YSZ suspension, N is the number of YSZ particles in the consolidated cake, and Z is the coordination number of YSZ particles in the cake. In the present analysis the coordination number was assumed to be 12 in a random close packing model. The measured packing density, E_c and W values were

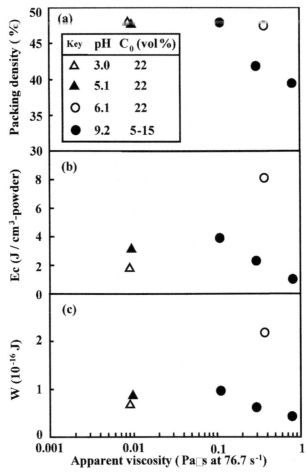

Fig. 7   Packing density (a), consolidation energy (Ec) for 1 cm^3 of YSZ particles (b) and energy applied between two particles (W) during consolidation (c) as a function of apparent viscosity of the initial suspension.

plotted in Fig. 7 as a function of apparent viscosity of the starting suspensions. The packing density showed a tendency to decrease as the viscosity of the suspensions increased. The consolidation energy, E_c for a similar solid content, also increased when the viscosity of the suspension increased. The decrease of solid content of the viscous suspension at pH 9.2 lowered the E_c value. The elimination of the solution through the porous cake at pH 9.2 needed no significant energy. In the suspension at pH 6.1, the relatively high E_c value was measured and packing density was higher than that for the suspension at pH 9.2. The tendency of Ec is reflected in the W value. The W values were

0.79×10^{-16}, 0.97×10^{-16} and 2.31×10^{-16} J for the suspensions at pH 3.0, 5.1 and 6.1. The W values of suspensions of 5-15 vol% at pH 9.2 were $0.59 \times 10^{-16} - 1.04 \times 10^{-16}$ J. As compared with the W values for the low viscosity suspensions at pH 3.0 and 5.1, the W values for the viscous suspensions at pH 6.1 became higher. This result suggests that the forming of agglomerated particle clusters into a high density cake needs a higher energy as compared with the forming of well dispersed particles. The calculated W values were 1000 times as high as the interaction energy between YSZ particles.

CONCLUSIONS

The pressure and energy required to consolidate an aqueous suspension with 67.8 nm YSZ particles at pH 3.0-9.2 was continuously measured using a newly developed pressure filtration apparatus. The viscosity of the YSZ suspension (isoelectric point, pH 2.8) was low at pH 3.0 and 5.1, and increased at pH 6.1 and 9.2. The consolidation of a 22 vol% suspension to 48 vol% solid required a low pressure (2.5 MPa) at pH 3.0 and 5.0. The suspension at pH 6.1 required a higher pressure (5 MPa) for consolidation. Further consolidation of the YSZ suspension was accompanied by the rapid increase of the applied pressure, and the packing density reached 50-53 % solids at 19 MPa. The decrease of the solid content in the viscous suspension at pH 9.2 increased the packing density. The forming of agglomerated particle clusters into a high density cake required that a higher consolidation energy is applied between two particles as compared with the forming of more dispersed particles. The height of YSZ powder compact under 19 MPa of applied pressure increased when the pressure was released, indicating the relaxation of stored elastic strain energy in the compressed YSZ compact.

REFERENCES

(1) I. A. Aksay, F. F. Lange, and B. I. Davis, Uniformity of Al_2O_3-ZrO_2 Composites by Colloidal Filtration, J. Am. Ceram. Soc., 66(10), C-190-C-192(1983).

(2) F. F. Lange, B. I. Davis, and E. Wright, Processing-related Fracture Origins: IV, Elimination of Voids Produced by Organic Inclusions, J. Am. Ceram. Soc., 69(1), 66-69(1986).

(3) L. M. Sheppard, International Trends in Powder Technology, Am. Ceram. Soc. Bull., 68(5), 979-985(1989).

(4) M. D. Sacks, H. W. Lee and O. E. Rojas, Suspension Processing of Al_2O_3/SiC Whisker Composites, J. Am. Ceram. Soc., 71(5), 370-379(1988).

(5) F. F. Lange, Powder Processing Science and Technology for Increased Reliability, J. Am. Ceram. Soc., 72(1), 3-15(1989).

(6) Y. Hirata, Theoretical Aspects of Colloidal Processing, Ceram. Inter., 23, 93-98(1997).

(7) Y. Hirata, I. A. Aksay and R. Kikuchi, Quantitative Analysis of Hierarchical Pores in Powder Compact, J. Ceram. Soc. Japan., 98(2), 126-135(1990).

(8) F. F. Lange and K. T. Miller, Pressure Filtration: Consolidation Kinetics and Mechanics, Am. Ceram. Soc. Bull., 66(10), 1498-1504(1987).

(9) C. H. Schilling, W. H. Shih and I. A. Aksay, Advances in the Drained Shaping of Ceramics, pp.307-320, Ceramic Transactions Vol.22, Ceramic Powder Science IV. Eds. S. Hirano, G. L. Messing and H. Hausner, Am. Ceram. Soc., Westerville, Ohio, 1991.

(10) F. F. Lange, New Interparticle Potential Paradigm for Advanced Powder Processing, pp.185-201 in Ceramic Transactions Vol.22, Ceramic Powder Science IV. Eds. S. Hirano, G. L. Messing and H. Hausner, Am. Ceram. Soc., Westerville, Ohio, 1991.

(11) H. J. Richter, Pressure Slip Casting of Silicon Nitride, pp.439-443 in Ceramic Transactions Vol.51, Processing Science and Technology, Eds. H. Hausner, G. L. Messing, and S. Hirano, Am. Ceram. Soc., Westerville, Ohio, 1995.

(12) R. A. Swalin, Thermodynamics of Solid. John Wiley & Sons, New York, 1972, p.6.

(13) Y. Hirata, S. Nakagawa and Y. Ishihara, Calculation of Interaction Energy and Phase Diagram for Colloidal Systems, J. Ceram. Soc. Japan., 98(4), 316-321(1990).

(14) J. A. Lewis, Colloidal Processing of Ceramics, J. Am. Ceram. Soc., 83(10), 2341-59 (2000).

(15) H. A. Barness, J. F. Hutton and K. Walters, An Introduction to Rheology, Elsevier Science Publishers, Amsterdam, The Netherlands, 1989, p.118.

(16) Y. Fukuda, T. Togashi, M. Naitou and H. Kamiya, Analysis of Electrosteric Interaction of Different Counter-ion Densities Using an Atomic Force Microscope, J. Ceram. Soc. Japan., 109(6), 516-520(2001).

(17) Y. Hirata, X. H. Wang, Y. Hatate and K. Ijichi, Electrical Properties, Rheology and Packing Density of Colloidal Alumina Particles, J. Ceram. Soc. Japan, 111(4), 232-237 (2003).

(18) C. F. Baes Jr and R. E. Mesmer, The Hydrolysis of Cations, Robert E. Krieger Publishing Company, Florida, 1986, p.159.

(19) R. A. Swalin, Thermodynamics of Solid. John Wiley & Sons, New York, 1972, p.233

COLLOIDAL PROCESSING AND SINTERING OF NANO-ZrO$_2$ POWDERS USING POLYETHYLENIMINE (PEI)

Yuji Hotta[1*], Cihangir Duran[1,2], Kimiyasu Sato[1] and Koji Watari[1]

[1]National Institute of Advanced Industrial Science and Technology,
Advanced Sintering Technology Group, Advanced Manufacturing Research Institute,
Anagahora 2266-98, Shimoshidami, Moriyama-ku, Nagoya, Japan
[2]Gebze Institute of Technology, Department of Materials Science and Engineering,
PK 141, 41400, Gebze, Kocaeli, Turkey
[*]Corresponding author

ABSTRACT

A stable colloidal suspension is important for fabricating dense samples with uniform microstructure using colloidal processing methods. Aqueous nano-ZrO$_2$ suspensions were prepared using polyethylenimine (PEI) as a dispersant. PEI adsorption on nano-ZrO$_2$ surfaces was influenced by PEI content and suspension pH. The isoelectric point (IEP) shifts from pH 7 at 0 wt% PEI to pH 10.3 at 3 wt% PEI. Stable suspensions had mean particle sizes in the range of 100 to 150 nm and sedimentation rates less than 0.4 mm/h, as compared to 2-5.5 μm and 10-50 mm/h for unstable suspensions. Electrostatic interactions, hydrogen bonding and PEI conformation were found to be controlling mechanisms on the colloidal stability of the suspensions. The amount of PEI adsorbed on nano-ZrO$_2$ surfaces was characterized using Thermogravimetric analysis (TG) and Fourier transform infrared spectrometer (FT-IR). The densification behavior of samples containing 3 wt% PEI at pH 7.1 sintered for 4 h at 1000 to 1300 °C was characterized. Relative density was found to increase rapidly from 54 % at 1100 °C to 92 % at 1200 °C and finally to 98 % at 1300 °C. Therefore, pellets were sintered at 1300 °C for 4 h to quantitatively correlate the processing conditions such as effect of pH and PEI content with densification.

INTRODUCTION

The dispersion of ceramic powders in liquid is of importance in the colloidal processing methods such as slip casting. These methods have been shown to be superior to conventional dry pressing in terms of controlling density and uniform microstructure evolution in the green and sintered states.[1, 2] If aggregates form, suspension stability and subsequently sintered properties are severely degraded. Therefore, aggregation must be prevented during colloidal processing[3].

Nano-particles can form aggregates in solution due to van der Waals attractive forces at short interaction distances.[4]. Therefore, sufficiently large stabilizing forces such as electrical double-layer repulsion or steric interactions should be used to create an energy barrier to inhibit aggregation.

Polyethylenimine (PEI) has been used as a dispersant for various ceramic powders and has been shown to enhance stability of ceramic powders in water[4-7]. When PEI is dissolved in a neutral or acidic solution, proton adsorption results in the protonation of the amine groups. Therefore, positively charged PEI easily adsorbs on the negatively charged ceramic surfaces, which provides an electrosteric effect that prevents aggregation of the ceramic powders[5].

In this study, the effects of PEI content and pH on the dispersion of nano-ZrO$_2$ powders in water were characterized. The objectives were to investigate both PEI adsorption mechanisms on nano-ZrO$_2$ particles and to define the pH range in which colloidally stable nano-ZrO$_2$ suspensions can be prepared to fabricate dense sintered ceramics. The effects of pH and PEI content on the suspension properties were characterized by measuring zeta potential, particle size and sedimentation rate. PEI adsorption on nano-ZrO2 powders was evaluated from thermogravimetric analysis (TG) and FT-IR measurements. Green samples were used to compare densification behavior at 1300°C as a function of pH and PEI content.

EXPERIMENTAL PROCEDURE

3 mol% yttria-stabilized nano-ZrO$_2$ powders with average particle size of 50 to 75 nm were purchased from Aldrich. Surface charge of nano-ZrO$_2$ particles was modified with polyethylenimine (PEI) (MW 10000, Anhydrous, Wako Pure Chemical Ind., Ltd., Japan). PEI was added at concentrations ranging from 0 to 3 wt% with respect to the dry weight of nano-ZrO$_2$.

1 vol% nano-ZrO$_2$ suspensions at various PEI contents (0 to 3 wt%) and pH values were prepared. First, PEI was dissolved in distilled water and then nano-ZrO$_2$ powder was added. Suspensions were ultrasonicated at 120 Watts for 10 min. pH was adjusted using reagent-grade HCl and NaOH (Wako Pure Chemical Ind., Ltd., Japan). Then, the suspensions were stirred for 6 h and finally centrifuged at 5000 rpm for 1 second to remove bigger particles from the suspension. Zeta potential was manually measured by applying 50 V (Model 502, Nihon Rufuto Co., Ltd, Japan). Particle size distribution was measured by using a laser particle size analyzer (Horiba LA-920, Japan). Sedimentation behavior of suspensions was characterized with a pulsed near infrared light (Turbiscan ma 2000, Formulaction, France). A clarified region at the top, and the sediment region at the bottom were characterized as a function of time. In sedimentation kinetics analysis, thickness of the clarification region from the top was considered and light transmitted at 3 % intensity was used.

The amount of PEI adsorption was determined from Thermogravimetric Analysis (TG) (Seiko Instruments SII, SSC/5200). Suspensions were centrifuged at 25000 rpm for 15 min to separate sediment and supernatant. The sediment was washed twice with distilled water to remove any excess (free) PEI. After final centrifugation, the sediment was dried under vacuum at 100°C for 2 h before analysis. TG of as-received ZrO$_2$ powder was chosen as a reference. Weight loss between 200-600°C was used in the adsorption calculations. The surface characterization of nano-ZrO$_2$ and PEI-modified nano-ZrO$_2$ powders was studied by FT-IR (Perkin Elmer, Spectrum GX, USA) after the samples were mixed with KBr powder. IR spectra of as-received PEI and 25 wt% aqueous PEI solutions at various pH values were measured after each solution was sandwiched as a thin layer between two CaF$_2$ plates.

Green bodies were fabricated by slip casting using gypsum molds. Densification behavior as a function of sintering temperature was first tested on the samples containing 3 wt% PEI prepared at pH 7.1 from 1000 to 1300 °C with an isothermal hold of 4 h. PEI burn out and sintering were performed at the same heating cycle as the pellets were heated at 100 °C/h. Densities were calculated using the Archimedes principle. The relative density was found to increase sharply from 54 % at 1100 °C to 92 % at 1200 °C and finally to 98 % at 1300 °C. Therefore, pellets were sintered at 1300 °C for 4 h to quantitatively correlate the processing conditions such as effect of pH and PEI content with densification. Microstructure observations were carried out by a scanning electron microscopy (SEM) (Model JSM5600N, JEOL, Japan).

RESULTS AND DISCUSSION

Figure 1 shows the effect of pH and PEI on zeta potential (ξ) of the nano-ZrO$_2$ suspensions. Electrophoresis properties of ZrO$_2$ in an aqueous solution determine the suspension stability. Attractive London-van der Waals forces should be overcome by repulsive forces such as electrostatic or polymeric to prepare stable suspensions[8]. In other words, high zeta potential value induces sufficiently high surface charge, which causes strong repulsive double-layer force[6]. Nano-ZrO$_2$ suspension with 0 wt% PEI has an isoelectric point (pH$_{iep}$) at nearly pH 7. PEI

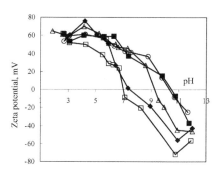

Figure 1 Zeta potential plots of suspensions as a function of pH and PEI content; ($\square$) 0 wt% PEI, ($\blacklozenge$) 0.5 wt% PEI, ($\triangle$) 1 wt% PEI, ($\blacksquare$) 2 wt% PEI, and ($\circ$) 3 wt% PEI.

interacts with nano-ZrO$_2$ surface and shifts the pH$_{iep}$ to more alkaline region as the amount of PEI content is increased. The pH$_{iep}$ increases from pH 7 at 0 wt% PEI to pH 10.4 at 3 wt% PEI. This can be attributed to the fact that PEI is a cationic polyelectrolyte and addition of a strong acid to the PEI-containing solution neutralizes -NH- basic groups, resulting in a positively charged polymer skeleton according to the following reaction[9];

$$-(CH_2-CH_2-NH-)_n + H_3O^+ \rightarrow -(CH_2-CH_2-NH_2^+-)_n + H_2O \qquad (1)$$

As a result, positively charged -NH$_2^+$- groups can adsorb on negatively charged nano-ZrO$_2$ surfaces by an electrostatic attraction, which causes the pH$_{iep}$ to shift to the alkaline region. Furthermore, nano-ZrO$_2$ powders have various negative zeta potentials in the alkaline region due to the different pH$_{iep}$ values such as ξ= -72 mV at 0 wt% PEI compared to ξ as low as -9 mV at 3 wt% PEI at pH 11. The zeta potential curves for the suspensions with PEI have a tendency to approximate the original curve of the suspension without PEI at highly acidic (e.g., pH 3) and highly alkaline (e.g., pH 12) solutions. It was reported that the degree of PEI dissociation (α) increases with decreasing pH, that is, fully dissociated PEI at pH 2 (α=1) and undissociated (e.g., free of charge) PEI

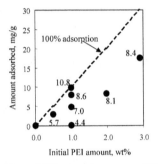

Figure 2 PEI adsorption on nano ZrO$_2$ powders as a function of initial PEI content and pH. Numbers indicate suspension pH.

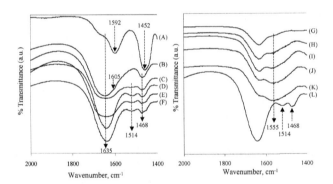

Figure 3. (a) FT-IR spectra of (A) as-received PEI and (B-F) aqueous PEI solutions at various pH; (B) pH 11.6, (C) pH 9.2, (D) pH 7.1, (E) pH 3.3 and (F) pH 0.4. (b) Effect of PEI content on FT-IR spectra of ZrO$_2$; (G) PEI-free ZrO$_2$, (H) 0.5 wt% PEI, (I) 1 wt% PEI, (J) 2 wt% PEI, (K) 3 wt% PEI and (L) PEI solution at pH 7.1.

structure after pH 11 (α=0) can be obtained. Therefore, PEI is strongly desorbed due to a strong electrostatic repulsion between the positively charged polymer and solid sites at pH 3, and PEI does not carry charge at pH 12.

Figure 2 shows the amount of PEI adsorbed on nano-ZrO$_2$ powders as a function of initial PEI content and pH. The plot represents the amount of PEI adsorbed per gram of nano-ZrO$_2$. The dashed line is given as a reference for the complete adsorption. The amount of PEI adsorbed is incomplete at the as-prepared pH conditions although it increases with increasing polymer content. The amount of PEI adsorbed increases steeply with increasing pH, that is, it is almost nil (e.g., 0.05 mg/g) at pH 4.4 in contrast to nearly full (9.9 mg/g) at pH 10.8.

Figures 3a and 3b compare the adsorption curves as a function of initial PEI contents and pH at 1 wt% PEI, respectively. The spectra of PEI-free ZrO$_2$ and the PEI solution at pH 7.1 are shown for reference in both graphs. Consistent with the TG data given in Figure 2, the broad absorption band at 1555 cm^{-1}, which becomes more dominant with increasing PEI content and pH in both plots, is evidence of the PEI adsorption onto the nano-ZrO$_2$ particles. Furthermore, a shift in the band position from 1514 cm^{-1} to 1555 cm^{-1} can be attributed to the bonding of PEI on nano-ZrO$_2$ surface via -NH$_2^+$- groups, as compared to the free PEI present in the solution.

Furthermore, TG and FT-IR data (Figs. 2, 3b, and 3c) show that the amount of PEI adsorbed increases with increasing initial PEI content and suspension pH. The adsorption

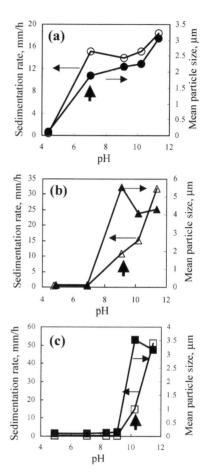

Figure 4 Mean particle size and sedimentation rate as a function of pH and PEI content; (a) 0 wt% PEI, (b) 1 wt% PEI and (c) 3 wt% PEI. Vertical arrows show pH$_{iep}$.

process can be explained by considering α and zeta potential (Fig. 1) as a function of pH. The sign and magnitude of electrical surface charge of nano-ZrO$_2$ in the suspension can be modified with pH, which, in turn, strongly affects the adsorption process. The main adsorption mechanism was reported to occur by hydrogen bonding between isolated ZrOH groups on the ZrO$_2$ surface and the C-N- groups of PEI[4]. Electrostatic repulsion begins to fade towards pH$_{iep}$ due to the decreased numbers of positively charged surface sites on ZrO$_2$, as compared to the ones at pH 4.4, which contributes to the adsorption process. After pH$_{iep}$ 7, nano-ZrO$_2$ particles have negative charge (Fig. 1) and, therefore, the positively charged PEI can easily and strongly adsorb on the negatively charged nano-ZrO$_2$ surfaces.

Figure 4 summarizes and compares mean particle size and sedimentation rates as a function of pH and PEI content. The suspensions with various PEI contents exhibit similar behavior with pH, in that, both mean size and sedimentation rate increase sharply on reaching pH$_{iep}$ of the respective suspensions, as indicated by the vertical arrows on each plot. In general, the suspensions whose pH<pH$_{iep}$ have sedimentation rates (mean particle sizes) of 0.3-0.4 mm/h (100-150 nm) as compared to 10-50 mm/h (2-5.5 μm) for the suspensions whose pH≥pH$_{iep}$. Full sedimentation took place within 4 to 12 hours for the suspensions whose pH≥pH$_{iep}$. At the pH$_{iep}$, a colloidal instability commences for each suspension, which can be attributed to the zero zeta potential (Fig. 1). Therefore, electrostatic attraction between oppositely charged sites gives rise to the aggregation of powders, which results in larger mean particle size, a broader size distribution, and faster sedimentation rates.

As mentioned before, a strong repulsive force is generated between powders when a sufficiently high surface charge is induced, as measured by high zeta potential value. The ξ values for the suspensions with 0 and 1 wt% PEI at pH 11.4 are -64 mV and -45 mV, respectively (Fig. 1). Despite these high ξ values, the mean particle size is big and sedimentation rate is high at the same pH (Figs. 4a and b). This can be attributed to the chemical properties of nano-ZrO$_2$ surface. For 0 wt% PEI suspension, the surface charge of the nano-ZrO$_2$ particles is negative in the alkaline region due to the adsorption of OH$^-$ ions[10]. Interconnection of nano-ZrO$_2$ particles with each other takes place by hydrogen bonding between highly negatively charged nano-ZrO$_2$ particles and water molecules, resulting in bigger particle size and, thus, higher sedimentation rate (Fig. 4a). As for the 1 wt% PEI-added suspension (Fig. 4b), bigger size and higher rate may arise from pH-dependent PEI behavior in the suspension. The α is very low in the alkaline region (e.g., ~ 0.1 at pH 10), which means that the PEI molecules adopt a compact coiled conformation [7]. As explained in the Section B, the polymer adsorption on nano-ZrO$_2$ surface is almost full at pH 10.8 (Fig. 3). Therefore, the hydrophobic and hydrogen interaction mechanisms between the PEI molecules, already adsorbed on nano-ZrO$_2$ particles, are likely to stick together and, hence, cause a coagulation. The same situation is also valid for the suspension with 3 wt% PEI at pH 11.5 (Fig. 4c). In addition, this suspension is unstable due to its proximity to the pH$_{iep}$ (10.4) and low ξ (-19 mV). |ξ| < 20 mV is considered to be unstable suspension[10]. These results indicate that suspension stability is deteriorated due to the interconnection of nano-ZrO$_2$ powders by hydrogen bonding and coiled conformation of PEI molecules in the alkaline region. In other words, the pH range where ξ is negative represents a kinetically unstable region for the current system, which is characterized by coagulation of the polymer-particle system. Figures 4b and c further indicates that the sedimentation rate increases after pH>pH$_{iep}$ even though the particle size decreases slightly. The particle size measurements were carried out right after the sample preparation step and then the suspensions were subjected to the sedimentation tests for 24 h. Aggregation of the particles with time takes place due to the instability of the suspensions. In this case, the gravitation force accelerates the settling of the aggregates. However, the stability range in nano-ZrO$_2$ suspensions is successfully extended up to pH 9.1 with 3 wt% PEI addition (Fig. 4c). This can be attributed to the adsorption and conformation of PEI in the region where the ξ is positive. The PEI molecules undergo a transition from a compact coiled to a stretched conformation with increasing α (or, alternatively, decreasing pH) due to an electrostatic repulsion between the neighboring ionized sites. Therefore, there happens a strong electrostatic repulsion between both positively charged PEI sites and nano-ZrO$_2$ powders at pH<~7, which in fact limits the adsorption but also prevents the coagulation of nano-ZrO$_2$ powders.

Figure 5 shows densification plots as a function of pH and PEI content. Arrows indicate the pH$_{iep}$ of the respective suspensions. Densification decreased suddenly on reaching pH$_{iep}$ for each

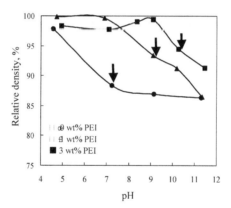

Figure 5 Densification plots as a function of pH and PEI content. Samples were sintered at 1300°C for 4 h. Arrows show pH$_{iep}$.

suspension. Irrespective of PEI content, the samples prepared from the suspensions whose pH<pH$_{iep}$ reached relative densities $\geq$ 98 % while densification was limited to $\leq$ 94% for the samples prepared from the suspensions whose pH$\geq$pH$_{iep}$. Figure 6 shows SEM micrographs of the

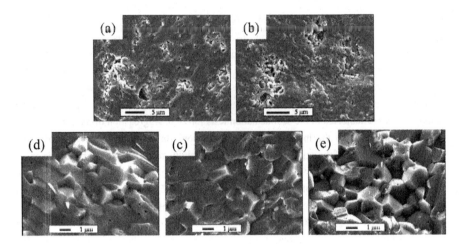

Figure 6 SEM images of fracture surfaces of sintered compacts prepared at various PEI contents and pH values; (a) 0 wt% PEI and pH 7.2, (b) 1 wt% PEI and pH 9.1, (c) 3 wt% PEI and pH 7.1, (d) 3 wt% PEI and pH 8.4, (e) 3 wt% PEI and pH 10.3, respectively.

fracture surfaces of the sintered samples. Figures 6a and 6b are micrographs of samples containing 0 and 1 wt% PEI prepared at their respective pH$_{iep}$ values, respectively. These samples are made of dense regions with less-dense regions in which large, micro-scale pores are present. This can be attributed to the formation of aggregates, due to the instability of suspension when the pH changes, that giving rise to differential sintering in the compacts. In contrast, samples containing 3 wt% PEI dispersed at up to pH 9.1 are almost fully dense (Figs. 6c and 6d). Thus, PEI broadens the pH range in which the stable suspensions can be prepared In turn, this leads to the fabrication of a dense sample. In the dense samples, mostly nano-scale pores were concentrated either along the grain boundaries or at the triple grain junctions which indicates an effective pore removal (or densification). Note that final grain size in the dense samples is in the range of 1 to 3 µm. A sample with 3 wt% PEI at pH$_{iep}$ 10.3 (Fig. 6e) shows microstructure evolution similar to that given in Figures 6a-b. These results strongly suggest that a dense ceramic can be fabricated from stable colloidal suspensions having a finer particle size and a narrow particle size distribution.

CONCLUSION

A colloidal dispersion of nano-ZrO$_2$ powders was successfully produced using polyethylenimine (PEI) as a dispersant. The addition of PEI shifted pH$_{iep}$ from pH 7 at 0 wt% PEI to pH 10.4 at 3 wt% PEI. Due to the electrosteric stabilization effect of PEI, the range for colloidal stability was extended up to pH 9.1. Electrostatic interactions, hydrogen bonding and PEI conformation were found to be the controlling mechanisms for the colloidal stability of the suspensions. The amount of PEI adsorbed on nano-ZrO$_2$ surfaces increased with increasing initial PEI content and suspension pH. Green bodies made from this colloidally stable nano-ZrO$_2$ suspension sintered compacts with 98 % of relative density on heating at 1300°C for 4 h.

REFERENCES

[1] I. A. Aksay, F.F. Lange and B.I. Davis, "Uniformity of Al$_2$O$_3$-ZrO$_2$ composites by colloidal filtration", *J. Am. Ceram. Soc.*, **66**, C-190 (1983).

[2] F.F. Lange, B.I. Davis and E. Wright, "Processing-related fracture origins: IV, elimination of voids produced by organic inclusions", *J. Am. Ceram. Soc.*, **69**, 66-68 (1986).

[3] V. A. Hackley, "Colloidal processing of silicon nitride with poly(acrylic acid): I, Adsorption and electrostatic interactions", *J. Am. Ceram. Soc.*, **80**, 2315-25 (1997).

[4] J. Wang and L. Gao, "Surface properties of polymer adsorbed zirconia nano particles", *NanoStructured Materials*, **11**, 451-57 (1999).

[5] A. Dietrich and A. Neubrand, "Effects of particle size and molecular weight of polyethylenimine on properties of nanoparticulate silicon dispersions", *J. Am. Ceram. Soc.*, **84**, 806-12 (2001).

[6] X. Zhu, T. Fengqiu, T. S. Suzuki and Y. Sakka, "Role of the initial degree of ionization of polyethylenimine in the dispersion of silicon carbide nanoparticles", *J. Am. Cer. Soc.*, **86**, 189-91 (2003).

[7] J. Wang and L. Gao, "Adsorption of polyethylenimine on nanosized zirconia particles in aqueous suspensions", *J. Coll. and Int. Sci.*, **216**, 436-39 (1999).

[8] R. Moreno, "The role of slip additives in tape-casting technology: Part I-solvents and dispersants", *Am. Ceram. Soc. Bull.*, **71**, 1521 (1992)

[9] S. Baklouti, C. Pagnoux, T. Chartier and J. F. Baumard, "Processing of aqueous α-Al$_2$O$_3$, α-SiO$_2$, α-SiC suspensions with polyelectrolytes", *J. Eur. Ceram. Soc.*, **17**, 1387-92 (1997).

[10] S. Vallar, D. Houivet, J. El Fallah, D. Kervadec and J.-M. Haussonne, "Oxide slurries stability and powders dispersion: Optimization with zeta potential and rheological measurements", *J. Eu. Ceram. Soc.*, **19**, 1017-21 (1999).

IMPORTANCE OF PRIMARY POWDER SELECTION IN AEROSOL DEPOSITION OF ALUMINUM NITRIDE

Atsushi Iwata, Jun Akedo
National Institute of Advanced Industrial Science and Technology (AIST)
Namiki 1-2-1
Tsukuba-shi, Ibaraki-ken, 305-8564 JAPAN

ABSTRACT

Aerosol deposition is a very promising method to make ceramic films several micrometers thick or thicker at room temperature from powder. In the case of aluminum nitride, films with hardness similar to conventionally sintered aluminum nitride plates were successfully deposited. Some of the very small crystallites in the deposited films transformed from the wurtzite crystal structure, which is stable in the ambient pressure, to the rock salt structure that is normally stable in a high pressure environment. Pre-processing, ball milling, and/or heat treatment of the primary (precursor) powder has been employed successfully to increase the film deposition rate. However, primary powders from different manufacturers show different results in deposition rate and in the crystal structure transformation. For the aluminum nitride powder made by company A, heat treatment was very effective, but heat treatment did not work well for the powder made by company B or C. Ball milling was effective for all powders. The transformation from wurtzite to rock salt occurs more in films made from Powder B that was both milled and heat treated, while only a small portion transforms in the films made by Powder A. Heat treated Powder C shows extensive transformation, while ball milled Powder C shows less transformation. Thus the selection of the primary powders is very important in aerosol deposition.

INTRODUCTION

Aerosol deposition is a very promising method to make ceramic films several micrometers thick or thicker. It is a process carried out at room temperature inside a reduced pressure chamber. No heating is necessary. Also no binders are necessary. A conventional ceramic powder is normally used as the primary powder in aerosol deposition. The primary powder is dispersed in a carrier gas inside the aerosol chamber, which forms an aerosol. The aerosol impinges on a substrate, driven by the pressure difference between the aerosol chamber and a vacuum evacuated deposition chamber. The aerosol, and hence, the primary ceramic powder particles, are accelerated through a nozzle before they impinge onto the substrate. The particles crash onto the substrate and form a solid film that firmly adheres to the substrate[1,2]. Many of the characteristics of a film deposited by aerosol deposition are similar to those of a thin plate made by conventional sintering. However the crystallite size of the films made by aerosol deposition is on the order of 10 nm, which is much smaller than in films prepared by conventional sintering. A wide range of ceramics, oxides, nitrides, and borides can be successfully deposited with this method. The substrates can be metals, glasses, ceramics and some plastics.

For the case of aerosol deposition of aluminum nitride, an interesting fact was found. The crystal structure of aluminum nitride that is stable at ambient pressure is wurtzite (hexagonal), and hence the starting primary aluminum nitride powder consists of wurtzite crystals. Aluminum nitride has two other crystal structures, rock salt (cubic) and zinc blende (cubic). These structures are stable at much higher pressure, such as 20 GPa and higher. Even though the primary aluminum nitride powder has only the wurtzite structure, most of the films deposited by aerosol deposition have both wurtzite crystallites and rock salt crystallites. This was confirmed through X-ray diffraction and transmission electron microscopy. These results indicate that the crystal structure of some of the original particles transforms from wurtzite to rock salt during aerosol deposition.

In aerosol deposition, the crystallite size in the deposited layer is much smaller than the particle size of the primary powder. In the aluminum nitride, the crystallite size of the deposited films is less than 100 nm, while the size of the primary powder is around 1 μm. Within the smaller crystallites in the films, only the crystallites less than ~20 nm transformed to rock salt. As the crystal size becomes small, the influence of surface energy on the formation of the stable crystal structure becomes significant. Therefore, the tendency toward a minimum surface to volume ratio increases, and the crystal structure tends to become more isotropic. The transformation from wurtzite to rock salt is believed to increase the isotropy of the crystal structure. Therefore the transformation occurs preferentially in the smaller crystallites. Also, the amount of rock salt crystallites in the deposited films depends on the aerosol deposition processing conditions[3].

Aluminum nitride is used as electronic substrates, heat sinks, and electronic packaging material, because it has high thermal conductivity, thermal expansion similar to that of silicon, and good dielectrical strength. The thickness of the aluminum nitride films successfully deposited by aerosol deposition is less than 30 μm. This thickness is insufficient for the above mentioned usage, so a thicker film and a faster deposition rate are required to make aluminum nitride by aerosol deposition. To achieve a faster deposition rate, pre-processing of the primary powder is said to be effective[2]. So far, ball milling and/or heat treating have been used.

The fact that the pre-processing of the primary powder is effective in accelerating the deposition means the characteristics of the primary powder affect aerosol deposition. Then a question arises. There are many manufacturers in the world who make aluminum nitride powder. The methods by which they make their powders are different, and the sizes and morphologies of the powders are also different. Then, do the differences between these powders make a difference in aerosol deposition?

No studies have been conducted to compare different aluminum nitride powders. Therefore the effects of powder pre-processing, gas flow rate, and the kind of primary powder on deposition rate and cubic/hexagonal ratio in aerosol deposited films was studied. Three different powders were examined. Aluminum nitride powder A was purchased from Alfa Caesar, powder B was made by Tokuyama Corporation, and powder C was made by Toyo Aluminum KK.

EXPERIMENTAL CONDITIONS
AlN films were fabricated by aerosol deposition using a machine developed in our laboratory. The flow rate of the Helium carrier gas ranged from 2 l/min to 30 l/min, and is controlled by a mass flow controller. The nozzle that accelerates the particles has a10 mm by 0.4 mm opening. The aerosol flows upwards from the nozzle. The vacuum pressure of the chamber varies with the He gas flow rate, and is on the order of 100 to several thousand Pa during the deposition.

The substrate is a 1.3 mm thick plate of soda-lime glass, known as "slide glass." It is placed under the XY stage. The distance between the nozzle opening and the substrate is set to 10 mm. The substrate is moved in a reciprocating motion of 10 mm amplitude at the velocity of 1.2 mm/s. With the 10 mm nozzle opening width and 10 mm displacement, the resulting size of the deposited aluminum nitride films are around 10 mm by 10 mm.

Three primary aluminum nitride powders are used. Powder A is Alpha Aesar aluminum nitride, N 32.0% min. powder. The nominal size of powder A is -325 mesh, which means the size is typically less than 4 μm. Powder B is Tokuyama aluminum nitride powder F grade. It has a nominal average particle size of 1.29 μm. Powder C is Toyo Aluminum powder JC grade. Its nominal average particle size is 0.8 μm. SEM photographs of the three powders are shown in Figures 1-3. All of these photographs are on the same scale. Powders B and C are smaller in size, and have smoother surface than powder A. These powders are used either as purchased, heated at 800□ for 4h in air, or ball milled in a Fritch P-5 ball milling machine at 400 rpm for 1, 3, 5 and 7 hours with zirconia balls.

The film thickness after 2 min of deposition is used as the index of the deposition rate. The deposition rate for the 10mm by 10mm area was determined by dividing the thickness by 2 min. However the deposition rate is not constant for long time, so this value should be deemed as the early average deposition rate.

The film thickness is measured with Tokyo Seimitsu diamond stylus profilometer, Surfcom 480A. The profile is taken over a 20mm length of the deposited film, and the average step height is calculated. Because of the residual stress inside the films, the substrates and the films have some curvature. This was compensated for in the analysis. One example of the sectional profile is shown in Figure 4. The film has a rougher surface than the glass substrate, and the thickness is not even, so an average values is used.

Figure 1.   SEM Photograph of as-received aluminum nitride powder A.

Figure 2. SEM Photograph of as-received aluminum nitride powder B.

Figure 3. SEM Photograph of as-received aluminum nitride powder C.

Standard X ray diffraction measurements using Cu Kα radiation were performed with a Rigaku RINT 2100V/PC diffractometer. As an index for the amount of rock salt aluminum nitride crystallites in the films deposited by aerosol deposition, the ratio of the strongest peak heights of rock salt and wurtzite in XRD patterns were calculated[4]. The strongest peak for rock salt aluminum nitride lies around a 2θ value of 44°, while the wurtzite aluminum nitride peak lies around 33°, as shown in Figure 5.

RESULTS AND DISCUSSION

Aluminum nitride films were successfully produced with all three primary powders by aerosol deposition. Figure 6 shows an example of a deposited film. This film was made from powder B, and has an average thickness of 5 μm. The colors of the films vary between black and white with a slight amount of yellow.

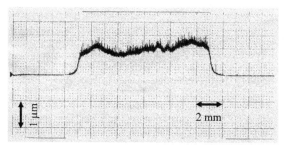

Figure 4   An example of the film thickness profile after compensating for the curvature.   The average thickness of this film is calculated to be 1.0 μm.

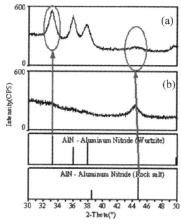

Figure 5. X-ray diffraction patterns of aluminum nitride films made by aerosol deposition.   There are fewer cubic crystallites in (a) than in (b).

Figure 7(a) shows the relationship between ball milling time and average film thickness for powder A after deposition for 2 min. A ball milling time of 0 signifies that the original powder was used without ball milling. Only thin films were deposited with the original powder A at any flow rate. However the thickness is slightly larger than that produced with powder B. Even at the 2l/min flow rate, ball milling increased the deposition rate. One hour ball milling increases the film thickness roughly 5 times. The film thickness seems to decrease with milling time greater than one hour in this experimental range.

Figure 7(b) shows the relationship between ball milling time and average film thickness for powder B. Only very thin films are deposited at any flow rate with the original powder. At a flow rate of 2 l/min most of the milled powders deposit only thin films. Much thicker films are produced using a flow rate of 6l/min. However the higher flow rates do not effectively increase the film thickness. Longer ball milling time up to 3h results in a thicker film, after which thickness stays constant with increasing milling time.

Figure 7(c) shows the relationship between ball milling time and average film thickness for powder C. As observed for powders A and B, ball milling increases the deposition rate, but long milling does not work effectively. At a flow rate of 2 l/min, all of the milled powders deposit only thin films. The deposition rate with powder C is slightly higher than with powder A or B over all of the ball milling times, and for the as-received powders. The deposition rate of as-received powder C is about ten times larger than that of powder B, although the absolute difference in film thickness is small.

There could be two reasons why ball milling increases the deposition rate. One may be that defects are introduced into the particles during ball milling. Particles are broken during aerosol deposition, and defects may help the particles break when impinging on the substrate. Perhaps the deposition rate increases as the number of broken particles increases, but it is not understood why. Another possible reason may be due to a change in particle size. From practical experience it is known that an optimal particle size range exists for deposition, and that is in the sub micrometer range. The original particles A and B are larger than 1 μm, and powder C is just below 1 μm. So, a higher deposition rate may be expected if the milled powder has a little smaller particle size due to the ball milling. However a difference in particle size was not clearly observed in SEM photographs of original and milled powders, and we have not found a reliable particle size measurement method to characterize the powders in this size range.

The effect of heating the primary powder as a pre-processing step is shown in Figure 8.

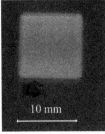

Figure 6. An Example of an aluminum nitride film made
by aerosol deposition. Thickness is 5 μm.

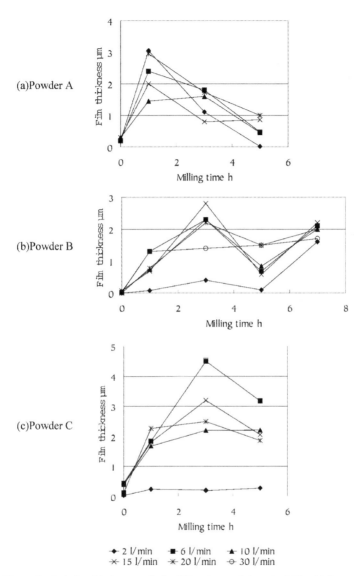

Figure 7. Relationship between ball milling time and deposited film thickness after 2 min. for the three different AlN powders.

The heat treatment results in a big difference between powders A, B, and C. In powder B and C, heating has almost no effect. The film thicknesses are the same as those of the as received powders at any gas flow rate. Powder A shows a definite change. Heat treating effectively increases the deposition rate. Also the rate of deposition increases as the He carrier gas flow rate increases. Films made from powder A are much thicker than the films made from ball milled powder as you can see from Figure 7(a). In Figure 8, the average film measured thicknesses is plotted for heat treated powder A to show the difference. The variation is in deposited film thickness at any given flow rate is very large, which indicates that the critical processing parameters are not yet effectively controlled.

The ratio of cubic to hexagonal aluminum nitride crystallites in the films made by aerosol deposition are shown in Figure 9, which also depicts the differences between the three powders. The ratios of cubic to hexagonal crystallites in films made from powder A, both milled and heat treated, are less than one, which means that there is less rock salt. The ratios in the films made from powder B, both milled and heat treated, are larger than one, which means that there is more rock salt. The films made from powder C show ratios larger than 1 for the heat treated powders and ratios less than 1 for the ball milled powders.

The results show that characteristics of aerosol deposition of aluminum nitride are very much affected by the as-received powder, and by the pre-processing conditions. Therefore it is very important to select a suitable primary powder and suitable pre-processing procedures in aerosol deposition.

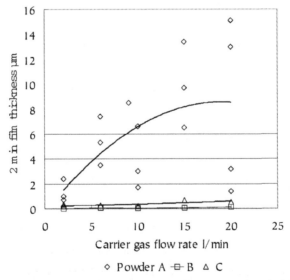

Figure 8. Relationship between carrier gas flow rate and deposited film thickness after 2 min for the three different heat treated AlN powders.

CONCLUSION

Aerosol depositions of aluminum nitride are carried out using three different powders. The as-received powders all have low deposition rate. Heat treating is effective in increasing the deposition rate with powder A, but is not effective with powders B or C. Ball milling is equally effective with all three powders. Powder B shows more transformation from wurtzite to rock salt easier, while powder A shows less. Heat treated powder C transforms easily, but milled powder C does not. The characteristics of the primary powder and the pre-processing conditions significantly affect aerosol deposition; thus, the selection of the primary powder and the pre-processing conditions is very important in aerosol deposition.

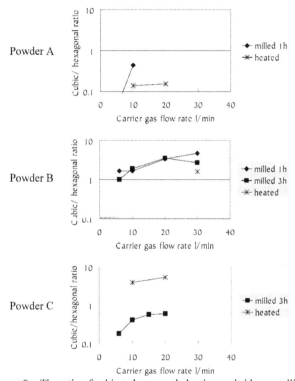

Figure 9. The ratio of cubic to hexagonal aluminum nitride crystallites in aerosol deposited films made from powders A, B, and C.

REFERENCES

[1]J. Akedo and M. Lebedev, "Microstructure and electrical properties of lead zirconate titanate (Pb(Zr-52/Ti-48)O-3) thick films deposited by aerosol deposition method," Jpn. J. Appl. Phys., 38, 5397-5401 (1999).

[2]J. Akedo, M. Lebedev, "Powder Preparation in Aerosol Deposition Method for Lead Zirconate Titanate Thick Films," Jpn. J. Appl. Phys., **41**, 6980 6984 (2002).

[3] A. Iwata, J. Akedo, M. Lebedev, "Cubic Aluminum Nitride Transformed Under Reduced Pressure Using Aerosol Deposition Method," J. Am. Ceram. Soc., **88**, 1067-1069 (2005).

[4]A Iwata, J Akedo, "Hexagonal to cubic crystal structure transformation during aerosol deposition of aluminum nitride," J. Cryst. Growth **275**, e1269–e1273 (2005).

PREPARATION OF 3D COLLOIDAL SPHERE ARRAYS USING BARIUM TITANATE FINE PARTICLES AND THEIR DIELECTRIC PROPERTIES

Satoshi Wada, Hiroaki Yasuno, Aki Yazawa, Takuya Hoshina, Hirofumi Kakemoto and Takaaki Tsurumi
Department of Metallurgy & Ceramics Science, Tokyo Institute of Technology,
2-12-1 Ookayama, Meguro, Tokyo 152-8552, Japan

ABSTRACT

Barium titanate ($BaTiO_3$) crystallites with various particle sizes from 22 to 500 nm were prepared by the 2-step thermal decomposition method of barium titanyl oxalate. Various characterizations revealed that these particles were impurity-free, defect-free, dense $BaTiO_3$ particles. The powder dielectric measurement clarified that the dielectric constant of $BaTiO_3$ particles with a size of around 58 nm exhibited a maximum of over 15,000. To explain this size dependence, the THz region dielectric properties of $BaTiO_3$ fine particles, especially Slater mode frequency, were measured using the far infrared (FIR) reflection method. As a sample for FIR measurement, the 3D colloidal sphere arrays were prepared using $BaTiO_3$ fine particles. Finally, the dense $BaTiO_3$ 3D colloidal sphere arrays were successfully prepared, and their dielectric properties were measured. As the result, the lowest Slater mode frequency was obtained at 58 nm. This tendency was completely consistent with particle size dependence of dielectric constant.

INTRODUCTION

Ferroelectric $BaTiO_3$ fine particles have been used as raw materials for electronic devices such as multilayered ceramic capacitors (MLCC). With the miniaturization of electronic devices, the down-sizing of MLCC has been developed. As a result, it is expected that the thickness of dielectric layers in MLCC will become less than 0.5 μm. Consequently, the particle size of the $BaTiO_3$ raw materials will decrease to a few ten nm. However, in ferroelectric fine particles, it was known that ferroelectricity decreases with decreasing particle and grain sizes, and disappears below certain critical sizes; this is called the "size effect" in ferroelectrics.[1-8] Therefore, the size effect in the $BaTiO_3$ is one of the most important phenomena for the industry and science.

To date, some researchers investigated the size effect of $BaTiO_3$ using particles.[3,4,6] Recently, Wada et al. reported a unique particle size dependence that dielectric maximum of 15,000 was observed at 68 nm.[9] This result suggested that the high dielectric constant of 15,000 should be originated from a change of dielectric polarization mechanism, i.e., (a) an appearance of new dipolar polarization at microwave region or (b) a change of the optic phonon behavior at THz region by down-sizing of the $BaTiO_3$ particles. Especially, among the optic phonon modes of the $BaTiO_3$, the lowest frequency mode, i.e., Slater mode, was most responsible optic phonon mode to determine large dielectric constant of the $BaTiO_3$. Therefore, it is very important to investigate the dielectric properties from sub-THz to THz regions for the $BaTiO_3$ particles[10]. However, there was no report to measure the size dependence of THz region dielectric properties for the fine particles. For this measurement, a new measurement method must be required.

In this study, the THz region dielectric properties of the $BaTiO_3$ fine particles were measured by using an IR reflection method. For this objective, the $BaTiO_3$ 3D colloidal sphere arrays were prepared using the $BaTiO_3$ fine particles. Moreover, the size dependence of the

Slater mode frequency was investigated.

EXPERIMENTAL DETAILS

Barium titanyl oxalate ($BaTiO(C_2O_4)_2 \cdot 4H_2O$) were prepared by Fuji Titanium Co., Ltd. Its Ba/Ti atomic ratio was 1.000 and the amount of the impurity was less than 0.02 mass%.[11]   To prepare defect-free, impurity-free $BaTiO_3$ nanoparticles, the 2-step thermal decomposition method[12] was used in this study.   As a result, $BaTiO_3$ particles with various particle sizes ranging from 22 to 500 nm were prepared.

The crystal structure was investigated using a powder X-ray diffractometer (XRD) (RINT2000, Rigaku, Cu-kα, 50 kV, 30 mA).   The average particle sizes and crystallite sizes were estimated using a transmission electron microscope (TEM) (CM300, Philips, 300 kV) and XRD.   The impurity in the products was analyzed using a Fourier transform infrared spectrometer (FT-IR) (SYSTEM 2000 FT-IR, Perkin Elmer) and by differential thermal analysis with thermogravimetry (TG-DTA) (TG-DTA2000, Mac Science).   The absolute density of the $BaTiO_3$ powders was measured using a pycnometer, and the relative density was calculated using a theoretical density estimated from lattice parameters by the XRD measurement.   The Ba/Ti atomic ratios for the $BaTiO_3$ particles were determined by using the X-ray fluorescence analysis. The dielectric constants for these $BaTiO_3$ particles were measured by using the powder dielectric measurement method[13].

For the FIR reflection measurement, 3D colloidal sphere array (colloidal crystal) was prepared using these $BaTiO_3$ fine particles from 22 to 500 nm.   A dense $BaTiO_3$ 3D colloidal sphere array was prepared from a $BaTiO_3$ slurry using a conventional method as shown in Fig. 1[10]. The FIR reflection spectra from 5 to 1,200 cm^{-1} were measured at 25 °C by using two FT-IR spectrometers, i.e., FARIS-1 (JASCO Co., 5 ~ 700 cm^{-1}) and FTIR-8600PC (Shimazu, 400 ~ 1,200 cm^{-1}).   Especially, the FIR reflection measurement using FARIS-1 was performed in vacuum to neglect the absorption peaks of H_2O in air.   A special attachment for a reflection measurement was used, and an incident and a reflection angles to the sample surface were fixed at 11°.   Aluminum evaporated film on the glass was used as a reference.   These two spectra were connected at 650 cm^{-1}.   The details of the measurement will be described later in this manuscript.

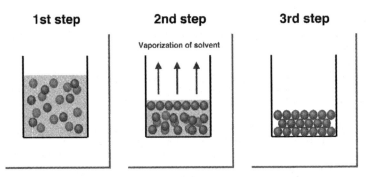

Fig. 1   Schematic diagram for a conventional preparation of the 3D colloidal sphere array.

RESULTS AND DISCUSSION

Preparation and characterization

$BaTiO_3$ fine particles with various particle sizes from 22 to 500 nm were prepared by the 2-step thermal decomposition method and characterized using various methods as shown in Table I. Investigation of impurity in these particles using both TG-DTA and FT-IR

Table I. Characterization results of the $BaTiO_3$ fine particles prepared using the 2-step thermal decomposition method.

2rd-step temperature	Particle size (nm)	Crystallite size (nm)	Impurity	Composition	Density
660 °C-1h	17	22	lattice: negligible surface: OH^-, CO_3^{2-}	Ba/Ti=1.00	5.78 g/cm^{-3} 97.3 %
760 °C-1h	33	34	lattice: negligible surface: OH^-, CO_3^{2-}	Ba/Ti=1.00	5.85 g/cm^{-3} 97.9 %
840 °C-1h	59	47	lattice: negligible surface: OH^-, CO_3^{2-}	Ba/Ti=1.00	5.90 g/cm^{-3} 98.6 %
860 °C-1h	68	58	lattice: negligible surface: OH^-, CO_3^{2-}	Ba/Ti=1.00	5.91 g/cm^{-3} 98.5 %
860 °C-1h*	102	98	lattice: negligible surface: OH^-, CO_3^{2-}	Ba/Ti=1.00	5.88 g/cm^{-3} 97.6 %
900 °C-1h	---	140	lattice: negligible surface: OH^-, CO_3^{2-}	Ba/Ti=1.00	5.89 g/cm^{-3} 98.6 %
960 °C-1h	---	230	lattice: negligible surface: OH^-, CO_3^{2-}	Ba/Ti=1.00	5.92 g/cm^{-3} 98.7 %
BT-05	500	500	lattice: negligible surface: OH^-, CO_3^{2-}	Ba/Ti=1.00	5.94 g/cm^{-3} 98.9 %

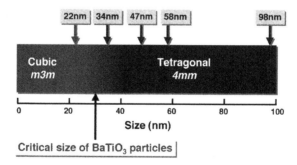

Fig. 2   Schematic diagram of relationship between crystal symmetry and particle size.

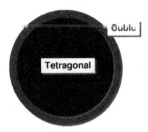

Fig. 3   Schematic diagram of a 2-phases model for the BaTiO₃ fine particles.

measurements revealed that no impurity was detected in the BaTiO₃ lattice while hydroxyl and carbonate groups were detected only on the surface.   Moreover, their relative densities were always over 97 % despite particle sizes.   Therefore, in this study, the impurity-free, defect-free, dense BaTiO₃ fine particles from 22 to 500 nm were successfully prepared.   Moreover, the crystal structure of these BaTiO₃ particles over 30 nm was assigned to tetragonal *4mm* by XRD measurement while that of the BaTiO₃ particles below 30 nm was assigned to cubic *m-3m* as shown in Fig. 2.   On the other hand, Raman scattering measurement revealed that all of BaTiO₃ particles prepared in this study were assigned to tetragonal *4mm* symmetry.

Size dependence of crystal structure
        The crystal structures of these BaTiO₃ fine particles were refined using a Rietveld method.   As results, it was clarified that all of the BaTiO₃ fine particles were composed of two parts, i.e., (a) surface cubic layer and (b) bulk tetragonal layer as shown in Fig. 3.   The structure refinement   revealed   that the thickness of the BaTiO₃ fine particles prepared in this study was around 1-2 nm.   This thickness was much smaller than that of commercial BaTiO₃ fine particles reported by Aoyagi *et al.*[14]   The size dependence of lattice parameters and tetragonality (*c/a*

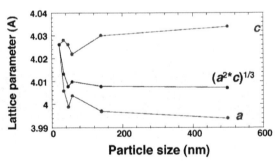

Fig. 4   Particle size dependence of lattice parameters and cell volume for the bulk tetragonal layer of the BaTiO₃ fine particles.

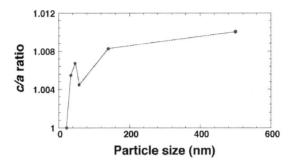

Fig. 5   Particle size dependence of tetragonality for the bulk tetragonal layer of the BaTiO₃ fine particles

ratio) in the bulk layer were also calculated as shown in Figs. 4 and 5.   As mentioned before, the critical size at the size-induced phase transition was around 30 nm.   However, despite the tetragonal *4mm* phase, the discontinuous change for the lattice parameter and tetragonality was clearly observed at 58 nm.   At present, it is difficult to explain this origin, but we believe that this discontinuous change should be related to the existence of the surface cubic layer.

Size dependence of dielectric property
        The dielectric constants of these particles were measured using suspensions by the powder dielectric measurement method.   Figure 6 shows particle size dependence of dielectric constants measured at 20.00 °C and 20 MHz.   With decreasing particle size, the dielectric constant was almost constant at around 3,000 down to 200 nm.   Below 200 nm, the dielectric constant increased with decreasing particle size, and at 58 nm, the dielectric maximum around 15,000 was clearly observed.   This result was almost consistent with the previous report[9].   Below 58 nm, the dielectric constant drastically decreased with decreasing particle size.   The results similar to this unique size dependence were reported for several BaTiO₃ ceramics[2,8] and particles[9,15].   Therefore, it is important to consider this origin of dielectric maximum at a certain

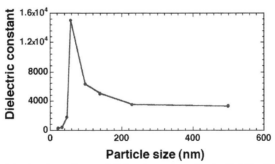

Fig. 6   Particle size dependence of dielectric constant for the BaTiO₃ fine particles

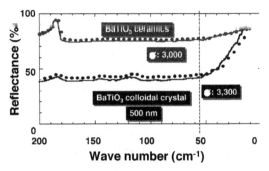

Fig. 7 FIR reflection spectra below 200 cm⁻¹ of the BaTiO₃ ceramics and the BaTiO₃ 3D colloidal sphere array.

size. In general, for BaTiO₃, it was known that the high dielectric constant over 3,000 can be originated from the phonon behavior, i.e., Slater mode frequency with the lowest frequency[16]. In this study, the dielectric maximum around 15,000 was observed at 58 nm. To explain this high dielectric constant, there is two hypothesis, i.e., (a) an appearance of new dipolar polarization at microwave region or (b) a softening of the optic Slater mode frequency. To confirm the origin, the THz-region dielectric properties must be measured for these BaTiO₃ fine particles.

Size dependence of phonon behavior

To explain this size dependence, the THz region dielectric properties of BaTiO₃ fine particles were measured using the FIR reflection method with the dense BaTiO₃ 3D colloidal sphere arrays. As a result, the high intensity reflection spectra below 100 cm⁻¹ for the BaTiO₃ fine particles were successfully obtained for the first time as shown in Fig. 8. As well known, the Slater mode was overdamped phonon mode, and the slope of the curve below 100 cm-1 increased with decreasing Slater transverse optic (TO) mode frequency as shown in Fig. 8.[17-18] Figure 9 shows the particle size dependence of FIR spectra below 100 cm⁻¹ measured at 25 °C.

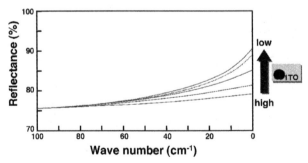

Fig. 8 Schematic relationship between Slater TO mode frequency and slope of the curve.

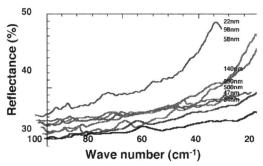

Fig. 9   Particle size dependence of FIR reflection spectra below 100 cm^{-1} for the BaTiO$_3$ fine particles.

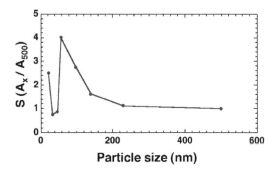

Fig. 10   Particle size dependence of normalized slope S of FIR reflection spectra below 100 cm^{-1} for the BaTiO$_3$ fine particles.

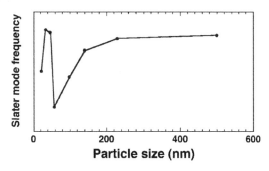

Fig. 11   Particle size dependence of relative values of the Slater TO mode frequency for the BaTiO$_3$ fine particles.

These slope was fitted using a 2nd order equation of $y = A^*x^2 + B^*x + C$ (A, B, C: constant), and we defined the slope of the curve using A value. Finally, all A values were divided by the A_{500} value obtained for the BaTiO$_3$ fine particles with 500 nm, and were normalized as $S = A/A_{500}$. Figure 10 shows the size dependence of the normalized slope S. Moreover, it should be noted that the reciprocal of the S value can be proportional to the Slater TO mode frequency.[17-18] However, in this study, owing the light scattering reflection, it was difficult to determine the absolute value for the Slater TO mode frequency. Thus, Fig. 11 shows the size dependence of the relative value obtained from Fig. 10 for the Slater TO mode frequency. Despite tetragonal *4mm* region over 30 nm, the minimum value for the Slater TO mode frequency was clearly observed at 58 nm, which revealed that the softening of the Slater TO mode occurred for the BaTiO$_3$ fine particles prepared in this study. Therefore, this result suggested that the high dielectric constant around 15,000 at 58 nm can be originated from the softening of the Slater TO mode.

CONCLUSION

The impurity-free, defect-free, dense BaTiO$_3$ fine particles were successfully prepared using the 2-step thermal decomposition method. It was confirmed that these particles intrinsically had the surface cubic layer with thicknesses around 1-2 nm. It is possible that this thin surface layer affected the whole crystal structure and phonon behavior of the BaTiO$_3$ fine particles. As a result, the dielectric maximum can be observed at 58 nm. This is just hypothesis, and much precise study must be required to explain this mechanism.

ACKNOWLEDGEMENT

We would like to thank Mr. M. Nishido of Fuji Titanium Co., Ltd. For preparing high purity barium titanyl oxalate. This study was partially supported by (1) a Grant-in-Aid for Scientific Research (**15360341**) from the Ministry of Education, Science, Sports and Culture, Japan and (2) the Ookura Kazuchika Memorial foundation.

REFERENCES

[1]K. Kinoshita, and A. Yamaji, "Grain-size Effects on Dielectric Properties in Barium Titanate Ceramics," *J. Appl. Phys.*, **45**, 371 (1976).

[2]G. Arlt, D. Hennings, and G. De With, "Dielectric Properties of Fine-grained Barium Titanate Ceramics," *J. Appl. Phys.*, **58**, 1619 (1985).

[3]K. Ishikawa, K. Yoshikawa, and N. Okada, "Size Effect on the Ferroelectric Phase Transition in PbTiO$_3$ Ultrafine Particles," *Phys. Rev. B*, **37**, 5852 (1988).

[4]K. Uchino, E. Sadanaga, and T. Hirose, "Dependence of the crystal structure on particle size in barium titanate," *J. Am. Ceram. Soc.*, **72**, 1555 (1989).

[5]M. H. Frey, and D. A. Payne, "Grain-size Effect on Structure and Phase Transformations for Barium Titanate," *Phys. Rev. B*, **54**, 3158 (1996).

[6]S. Wada, T. Suzuki, and T. Noma, "Role of lattice defects in the size effect of barium titanate fine particles: A new model," *J. Ceram. Soc. Jpn.*, **104**, 383 (1996).

[7]D. McCauley, R. E. Newnham, and C. A. Randall, "Intrinsic size effects in a BaTiO$_3$ glass ceramic," *J. Am. Ceram. Soc.*, **81**, 979 (1998).

[8]M. H. Frey, Z. Xu, P. Han and D. A. Payne, "The Role of Interfaces on an Apparent Grain Size Effect on the Dielectric Properties for Ferroelectric Barium Titanate Ceramics," *Ferroelectrics*, **206-207**, 337 (1998).

[9]S. Wada, H. Yasuno, T. Hoshina, S.-M. Nam, H. Kakemoto, and T. Tsurumi, "Preparation of nm-sized barium titanate fine particles and their powder dielectric properties," *Jpn. J. Appl. Phys.*, **42**, 6188 (2003).

[10]S. Wada, H. Yasuno, T. Hoshina, H. Kakemoto, Y. Kameshima, T. Tsurumi, and T. Shimada, "THz Region Dielectric Properties of Barium Titanate Fine Particles using Infrared Reflection Method," *J. Eur. Ceram. Soc.*, (2005) in press.

[11]T. Kajita, and M. Nishido, "Preparation of Submicron Barium Titanate by Oxalate Process," *Ext. Abst. 9th US-Japan Sem. Dielect. Piezo. Ceram.*, 425 (1999).

[12]S. Wada, M. Narahara, T. Hoshina, H. Kakemoto, and T. Tsurumi, "Preparation of nm-sized $BaTiO_3$ Fine Particles Using a New 2-step Thermal Decomposition of Barium Titanyl Oxalates," *J. Mater. Sci.*, **38**, 2655 (2003).

[13]S. Wada, T. Hoshina, H. Yasuno, S.-M. Nam, H. Kakemoto, and T. Tsurumi, "Preparation of nm-sized $BaTiO_3$ Crystallites by The 2-step Thermal Decomposition of Barium Titanyl Oxalate and Their Dielectric Properties," *Key Eng. Mater.*, **248**, 19 (2003).

[14]S. Aoyagi, Y. Kuroiwa, A. Sawada, I. Yamashita, and T. Atake, "Composite Structure of $BaTiO_3$ Nanoparticle Investigated by SR X-ray Diffraction," *J. Phys. Soc. Jpn.*, **71**, 1218 (2002).

[15]T. Hoshina, H. Yasuno, S.-M. Nam, H. Kakemoto, T. Tsurumi, and S. Wada, "*Size and Temperature Induced Phase Transition Behaviors of Barium Titanate Nanoparticles*," *Trans. Mater. Res. Soc. Jpn.*, **29**, 1207 (2004).

[16]T. Mitsui, I. Tatsuzaki, and E. Nakamura, *An Introduction to the Physics of Ferroelectrics*, vol. **1**, Gordon and Breach Science, New York, 1966.

[17]J. M. Ballantyne, "Frequency and Temperature Response of the Polarization of Barium Titanate," *Phys. Rev.*, **136**, A429 (1964).

[18]A. S. Barker, Jr., "Temperature Dependence of the Transverse and Longitudinal Optic Mode Frequencies and Charges in $SrTiO_3$ and $BaTiO_3$," *Phys. Rev.*, **145**, 391 (1966).

Sintering and Properties

SINTERING AND MECHANICAL PROPERTIES OF SiC USING NANOMETER-SIZE POWDER

Nobuhiro Hidaka* and Yoshihiro Hirata
Department of Advanced Nanostructured Materials Science and Technology,
Graduate School of Science and Engineering, Kagoshima University
1-21-40 Korimoto, Kagoshima 890-0065, Japan
* Research Fellow of the Japan Society for the Promotion of Science

ABSTRACT

To homogeneously distribute Al_2O_3 and yttrium ions as sintering additives in SiC, an 800 nm SiC powder (75-100 vol%) was mixed with a 30 nm SiC powder (0-25 vol%) in a pH 5, 0.05-0.1 M-$Y(NO_3)_3$ solution containing 0.2 μm median size Al_2O_3 powder. In the 20 vol% SiC suspension, a network structure of negatively charged SiC particles was formed by the heterocoagulation of the positively charged Al_2O_3 and yttrium ions. The apparent viscosity of the aqueous suspension of the SiC-Al_2O_3 (1.2 vol%)-Y^{3+} ions (0.94-1.13 vol% Y_2O_3)-polyacrylic acid (dispersant) system increased with increasing volume fraction of 30 nm SiC powder. The suspension was consolidated by casting in a gypsum mold to form a 36-40 % theoretical density green compact. The SiC compact was hot-pressed at 1950°C in Ar to 97.3-99.4 % relative density. The addition of 30 nm SiC to 800 nm SiC decreases the grain size and flaw size in the hot-pressed SiC, and improves the flexural strength. The flexural strength increases gradually with increasing 30 nm SiC fraction, and reaches a constant value at 15-25 vol% 30 nm SiC. No significant improvement in fracture toughness is measured with 30 nm SiC addition. Increasing the amount of sintering additives decreased both the strength and fracture toughness.

INTRODUCTION

Silicon carbide (SiC) is potentially useful as a high temperature structural material because of its high strength, high hardness, high creep resistance, and high oxidation resistance. However, it has been impossible to densify submicrometer-size SiC powder without sintering additives because of its strong covalent bonding character. Recently chemical methods of adding sintering additives such as Al_2O_3 and Y_2O_3 to SiC powder have been studied to control the liquid phase sintering and the resultant microstructure of SiC ceramics [1-11]. The densification of SiC occurs by a dissolution-reprecipitation mechanism involving solid SiC in a SiO_2-Al_2O_3-Y_2O_3 liquid phase [12-24].

SiO_2 forms naturally on the surface of as-received SiC powder. Consequently, chemical methods of introducing sintering additives to SiC are expected to provide the following advantages: 1) a homogeneous distribution of the sintering additives around the SiC particles, 2) an increased densification rate due to the well-distributed liquid, and 3) a decrease of the amount of additives required for densification [25].

In previous work [26,27], the interaction of silicon carbide with alumina (1.17 vol%) and yttrium ions (0.94 vol% Y_2O_3) in an aqueous suspension at pH 5.0 was investigated as a means to homogeneously disperse a small amount of the sintering additives in SiC powder. Both the SiC and Al_2O_3 were added in powder form, while the Y_2O_3 was added as yttrium ions in solution. Hot-pressing of a consolidated SiC compact at 1850°-1950°C produced a dense SiC ceramics (95-98 % of theoretical density) with the excellent mechanical properties. The effect of mixing in nanometer-size SiC powder on the processing of submicrometer-size SiC powder was also reported. The addition of 25 vol% of 30 nm SiC particles to the submicrometer-size SiC (800 nm, 75 vol%) improved the mechanical properties. The hot-pressed ceramics had an average four-point flexural strength of 812 MPa, a fracture toughness of 6.0 MPa·m$^{1/2}$, a Vickers hardness of 16-20 GPa, and a Weibull modulus 5.8 [28]. In this paper, the influence of the volume fraction of nanometer-size SiC particles on the processing, sinterability and mechanical properties of submicrometer-size SiC were studied. Green compacts were formed from aqueous SiC suspensions containing Al_2O_3 particles and Y^{3+} ions, and hot-pressed at 1950°C in Ar.

EXPERIMENTAL PROCEDURE

An α-SiC powder, supplied by Yakushima Electric Industry Co., Ltd., Kagoshima, Japan, (designated SiC A) was used. It had a nominal chemical composition of 98.90% (by mass) SiC, 0.66% SiO_2, 0.37% C, 0.004% Al, and 0.013% Fe. Its median size was 800 nm, and it had a specific surface area of 13.4 m^2/g. Another plasma CVD-processed SiC, supplied by Sumitomo Osaka Cement Co., Ltd., Tokyo, Japan (designated SiC B) was also used. Its chemical composition was 98.90% (by mass) SiC, 0.90% SiO_2, and 3.50 % C. Its median size was 30 nm, and it had a specific surface area of 50.9 m^2/g. A submicrometer–size α-Al_2O_3 powder (sintering additive) was mixed with the as–received SiC powders. It had a purity of >99.99 mass% Al_2O_3, a median size 200 nm, and a specific surface area 10.5 m^2 /g (Sumitomo Chemical Industry Co., Ltd., Tokyo, Japan). The zeta potential of the SiC A, SiC B, and α-Al_2O_3 was measured as a function of pH at a constant ionic strength of 0.01M- NH_4NO_3 (Rank Mark II, Rank Brothers Ltd., UK).

Six SiC suspensions were consolidated into a rectangular shape 17 mm in height, 25 mm in width, and 38 mm in length by casting in a gypsum mold. Thirty vol% SiC powder was dispersed in a pH 5.0, 0.3 M- $Y(NO_3)_3$ aqueous solution, and the α-Al_2O_3 powder was mixed in. The amount of yttrium ions adsorbed on the SiC particles or the Al_2O_3 particles was determined by chelatemetric titration of yttrium ions in the filtrate. The volume ratio of the SiC / Al_2O_3 / Y_2O_3 components was adjusted to 1 / 0.012 / 0.0094 (designated A)[27]. The SiC A and SiC B powders were mixed at a volume ratio of SiC A / SiC B = 95 / 5 (designated M-1), 90 / 10 (designated M-2), 85 / 15 (designated M-3) and 75 / 25 (designated M-4). They were dispersed with 20 vol% solids in 0.05-0.1 M- $Y(NO_3)_3$ aqueous solutions at pH 5.0. 2.25 mass% polyacrylic acid (PAA, average molecular 10000, Daiichi Kogyo Seiyaku Co., Kyoto, Japan), relative to the SiC powder, was added to the suspension. The α-Al_2O_3 powder was added to the SiC powder suspension at a volume ratio of SiC / Al_2O_3 / Y_2O_3 = 1 / 0.012 / 0.009-0.013. Sample M-5, with a volume ratio SiC (SiC A 75 %, SiC B 25 %) / Al_2O_3 / Y_2O_3 = 1 / 0.012 / 0.033 was also processed in a 0.3 M- $Y(NO_3)_3$ solution containing PAA.

Green SiC compacts were hot-pressed at 39 MPa and 1950 °C for 2 h in flowing Ar (High Multi 5000, Fujidempa Kogyo Co., Ltd., Osaka, Japan). The surface of hot-pressed SiC was etched with the mixture of NaCl / NaOH = 85/15 (molar ratio) to observe the microstructures by scanning electron microscopy (SM 300, Topcon Technologies, Inc., Tokyo, Japan). A strain gauge was attached to the tensile plane of specimen to measure the Young's modulus. The flexural strength of SiC polished with 1 µm diamond paste was measured at room temperature using the four-point flexural method [29] over spans of 30 mm (lower span) and 10 mm (upper span) using a crosshead speed of 0.5 mm/min. The fracture toughness was evaluated using a single-edge V-notch beam (SEVNB) method [27].

RESULTS AND DISCUSSION

Properties of the mixed SiC powder system suspension

The isoelectric point of the SiC A, SiC B and Al_2O_3 was pH 2.8, 2.8 and 7.7, respectively. No difference in the zeta potentials was measured between the SiC A and the SiC B. The surfaces of SiC particles were coated with thin SiO_2 films [3]. At a pH below the isoelectric point (pH 2.8), the number of positively charged $SiOH^{2+}$ sites becomes greater than the number of negatively charged SiO^- sites. The opposite occurs at a pH above the isoelectric point. Similary, the number of $AlOH^{2+}$ sites becomes greater than that of AlO^- sites at a pH below the isoelectric point. In previous experiments [28,30], it was found that yttrium ions adsorb on the surfaces of the negatively charged SiC A and SiC B at pH 5. No yttrium ions adsorb on the Al_2O_3 surface at pH 5.0. When the pH of the $Y(NO_3)_3$ solution is increased above 6.5, a cloudy precipitate of $Y(OH)_3$ forms. This result agrees with the previously reported solubility limit of $Y(OH)_3$ as a function of pH [31]. In this study, it was difficult to prepare a fluid 30 vol% powder suspension. Therefore, 2.2 mass% of polyacrylic acid was added to the mixed SiC suspension. PAA $[(- CH_2 - CH(COOH) -)_n]$ dissociates to produce negatively charged polymer and H^+ ions at pH higher than 3 [32]. The negatively charged polymer is adsorbed on the positively charged $SiOH^+$ sites of SiC surfaces. The addition of PAA

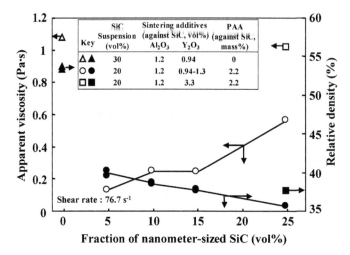

Fig.1 Influence of SiC B addition on the apparent viscosity of the SiC suspension at a shear rate of 76.7 s^{-1} and the density of SiC compacts consolidated.

decreases the apparent viscosity of the SiC suspension, indicating that the dispersibility of the SiC particles is increased by the adsorption of PAA.

Figure 1 shows the influence of the addition of SiC B on the apparent viscosity of the SiC suspension at a shear rate of 76.7 s^{-1}, and on the density of the consolidated SiC compacts. The apparent viscosity of the 20 vol% SiC mixed powder suspension is lower than that of a 30 vol% SiC A suspension because of the lower solid content and the electrosteric interaction of the adsorbed PAA. The apparent viscosity of the mixed powder suspension increases with the addition of SiC B, indicating the formation of the particle network by the heterocoagulation of the negatively charged SiC – positively charged Al_2O_3 – negatively charged PAA – Y^{3+} ion system. The packing density of the mixed powder compact is lower than that of the SiC A compact and decreases with increasing amount of SiC B.

Hot-pressing of SiC compacts

Figure 2 shows the relative density of the SiC during heating under a pressure of 39 MPa, determined from shrinkage curves. The densification of the A compact with green density of 53.3 % started at 1420°C, which is close to the liquid formation temperature (~1400°C) in the SiO_2-Al_2O_3-Y_2O_3 system [33]. Compared to the A compact, the M-1, M-3, and M-4 compacts began to shrink at a lower temperature. As seen in Fig.2, the densification of M-1, M-3 and M-4 compacts was enhanced even before the liquid formation temperature of 1420°C in the phase diagram [33].

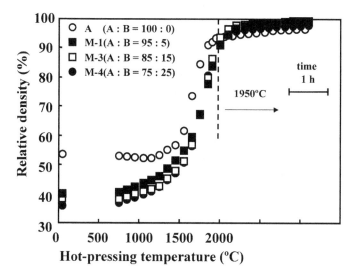

Fig. 2 Relative density of the SiC during heating under a pressure of 39 MPa, determined from the shrinkage curves

This result may possibly be related to the sintering of the oxide additives (Al_2O_3- Al_2O_3, Al_2O_3-Y_2O_3 or Y_2O_3-Y_2O_3) dispersed in the SiC matrix, including the nanometer-size SiC. Figure 3 shows the density of the SiC after hot-pressing at 1950°C as a function of the volume fraction of SiC B. The relative density reaches 97.3 ± 1.3 %, 99.2 ± 0.4 %, 99.3 ± 0.9, 99.4 ± 0.4 %, and 97.7 ± 1.5 % in the SiC with 0, 5, 10, 15 and 25 % SiC B, respectively. The addition of SiC B enhances the sinterability of SiC A. This result is explained by the fast dissolution and precipitation rate (densification rate) of the SiC B particles in the liquid formed. Additionally, the formation of closed pores is inhibited in the final stage of sintering. The porosity of closed pores was 0.2-0.6 % for the A compact and 0.02-0.2 % for the M-1 compact.

The role of the SiC B on densification is also important. As seen in Fig.3 (a), the sinterability of the SiC decreases at 25 vol% SiC B. This result is most likely due to the lack of sintering additives per m^2 of SiC surface. The specific surface area of the starting powder is larger for the SiC B than for the SiC A. Therefore, for a constant volume of sintering additives, the effective amount of the sintering additives per m^2 of SiC decreases with SiC B content as shown in Fig. 3(b). With the same 25vol% SiC B, the hot-pressed density increases to 99.7 % when the amount of Y_2O_3 is increased to 3.3 vol%. From the result in Fig. 3, the critical amount of sintering additive for full densification is estimated to be 2 mg (Al_2O_3 + Y_2O_3) per m^2 of SiC.

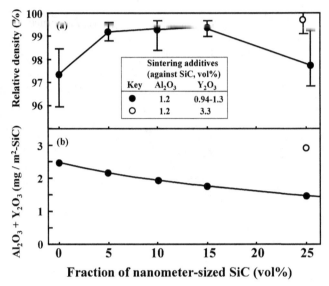

Fig. 3 Density (a) of SiC after the hot-pressing at 1950°C and amount of sintering additives per m² of SiC surface (b), as a function of SiC B.

Figure 4 shows the microstructures of SiC hot-pressed at 1950°C. The hot-pressed microstructure consists of equiaxed grains. The average grain size, measured on 200 grains, is: a) 3.5 μm for compact A; b) 2.5 μm for compact M-1; c) 2.3 μm for compact M-3; and d) 2.0 μm for compact M-4. The addition of SiC B to SiC A decreases the average grain size. The chemical interaction between the SiC particles and the oxide liquid increases with decreasing particle size. The solubility of SiC, [S]$_r$, with a particle radius, r, in the liquid at a given temperature (T) is expressed by Eq. (1) (The Thomson-Freundlich equation) [34],

$$\ln\frac{[S]_r}{[S]_\infty} = \frac{2M\sigma}{rRT\rho} \qquad (1)$$

where [S]$_\infty$ is the solubility of SiC at r → ∞, M is the molecular weight of SiC, σ is the interaction energy at the interface between the solid SiC and the liquid, R is the gas constant and ρ is the density of the SiC. Equation (1) indicates that the solubility of the SiC increases exponentially with decreasing particle size. Based on the above relationship, it is proposed that the nanometer-size SiC

Fig.4   Microstructures of SiC hot-pressed at 1950°C

particles dissolve faster in the liquid, and precipitate on the surfaces of   the submicrometer-size SiC particles. The above dissolution-reprecipitation mechanism proceeds during the hot-pressing.

Mechanical properties of hot-pressed SiC

Figure 5 shows the flexural strength (a) and fracture toughness (b) as a function of the volume fraction of SiC B. The mean failure strength and fracture toughness are 565 MPa and 5.4 MPa·m$^{1/2}$ in compact A, 744 MPa and 4.8 MPa·m$^{1/2}$ in compact M-1 (5 vol% SiC B), 805 MPa and 5.1 MPa·m$^{1/2}$ in compact M-3 (15 vol% SiC B) and 812 MPa and 6.0 MPa·m$^{1/2}$ in compact M-4 (25 vol% SiC B), respectively. In compacts M-3 and M-4, the maximum strength reaches 1069-1076 MPa. The addition of SiC B is effective in increasing the strength of the hot-pressed SiC. However, only a small improvement in the fracture toughness is measured with the SiC B addition.

The increased flexural strength with the addition of SiC B is due to a decrease in the flaw size [28]. Increasing the amount of sintering additives (M-5 sample) at 25 vol% SiC B is effective in enhancing the sinterability (Fig.3), but it decreases both the strength and fracture toughness. With increasing density, the flaw sizes observed in sample M-5 become smaller than those observed in sample M-4 with the same SiC B content [28]. Consequently, the decrease in the strength of M-5 sample is due to the decrease in the fracture toughness as shown in Fig.5 (b).

The fracture toughness (K_{IC}) of the SiC compact depends on Young's modulus (E) and fracture energy (γ) ($K_{IC} = (2E\gamma)^{1/2}$). Figure 6 shows the Young's modulus and fracture energy of

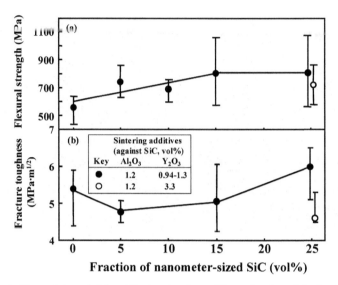

Fig.5 Flexural strength (a) and fracture toughness (b) as a function of the fraction of SiC B.

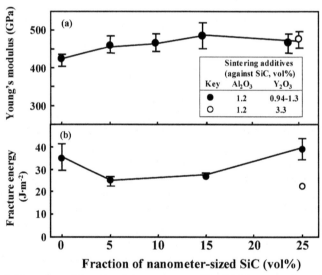

Fig.6 Young's modulus (a) and fracture energy (b) of SiC hot-pressed as a function of the fraction of SiC B.

hot-pressed SiC. The change in the Young's modulus with SiC B addition is similar to the change in density (Fig.3), reflecting the influence of porosity. The dependence of fracture energy on SiC B content is similar to the results in Fig.5(b). The fracture surface of hot-pressed SiC indicates that the cracks propagate along the grain boundaries. The fracture energy is related to the mechanical properties of grain boundaries. The toughening mechanisms in the microstructures shown in Fig.4 are grain bridging, crack branching and crack deflection. These mechanisms are influenced by the grain size. A decrease in grain size with the addition of SiC B is accompanied by a decrease in the grain bridging effect (effect 1), a decrease in crack deflection (effect 2) and an increase in crack branching effect (effect 3). These three effects affect the fracture toughness. The relative low fracture energy in sample M-5 with a higher amount of sintering additives as compared with sample M-4 may be associated with a weaker bond at the grain boundaries enriched with oxide additives.

CONCULSIONS

This paper reports on the influence of the addition of 30 nm SiC particles (SiC B) and sintering additives on the densification and mechanical properties of hot-pressed SiC formed from an aqueous suspension of 800 nm SiC particles (SiC A):

(1) In an aqueous suspension of negatively charged SiC, positively charged $Al_2O_3 - Y^{3+}$ ions, and negatively charged polyacrylic acid at pH 5, a heterocoagulated particle network was formed.

(2) Relative to SiC A alone, porous SiC compacts of mixtures of SiC A and SiC B start to densify at a lower temperature of 1000°C and sinter to 97.3-99.4 % theoretical density after hot-pressing at 1950°C. The addition of SiC B to SiC A increases the density and decreases the grain size and flaw size of the hot-pressed SiC.

(3) The addition of SiC B (5-25 vol%) is effective in increasing the flexural strength. However, no significant improvement in fracture toughness is measured with the addition of SiC B. Increasing the amount of sintering additives is effective for enhancing the sinterability, but decreases both the strength and fracture toughness.

REFERENCES

[1]E. Liden, E. Carlstrom, L. Eklund, B. Nyberg and R. Carlsson, "Homogeneous Distribution of Sintering Additives in Liquid-Phase Sintered Silicon Carbide," *J. Am. Ceram. Soc.*, **78[7]**, 1761-1768 (1995).

[2]Y. Hirata, K. Miyano, S. Sameshima and Y. Kamino, "Reaction between SiC Surface and Aqueous Solutions Containing Al Ions," *Colloid and Surfaces, A : Physicochem. & Eng. Aspects*, **133**, 183-189 (1998).

[3]Y. Hirata, K. Hidaka, H. Matsumura, Y. Fukushige and S. Sameshima, "Colloidal Processing and Mechanical Properties of Silicon Carbide with Alumina," *J. Mater. Res.*, **12**, 3146-3157 (1997).

[4]S. Sameshima, K. Miyano and Y. Hirata, "Sinterability of SiC Powder Coated Uniformly with Al Ions," *J. Mater. Res.*, **13**, 816-820 (1998).

[5]M. A. Mulla and V. D. Krstic, "Low-Temperature Pressureless Sintering of β-Silicon Carbide with Aluminum Oxide and Yttrium Oxide Additions," *Am. Ceram. Soc. Bull.*, **70**, 439-443 (1991).

[6]X. H. Wang and Y. Hirata, "Colloidal Processing and Mechanical Properties of SiC with Al_2O_3 and Y_2O_3," *J. Ceram. Soc. Japan.*, **112[1]**, 22-28 (2004).

[7]L. M. Wang and W. C. Wei, "Colloidal Processing and Liquid-Phase Sintering of SiC," *J. Ceram. Soc. Japan.*, **103[5]**, 434-443 (1995).

[8]Y. W. Kim, J. Y. Kim, S. H. Rhee and D. Y. Kim, "Effect of Initial Particle Size on Microstructure of Liquid-Phase Sintered α-Silicon Carbide," *J. Eur. Ceram. Soc.*, **20**, 945-949 (2000).

[9]V. V. Pujar, R. P. Jensen and N. P. Padture, "Densification of Liquid-Phase-Sintered Silicon Carbide," *J. Mater. Sci. Lett.*, **19**, 1011-1014 (2000).

[10]E. Liden, M. Persson, E. Carlstorm and R. Carlsson, "Electrostatics Adsorption of a Colloidal Sintering Agent on Silicon Nitride Particles," *J. Am. Ceram. Soc.*, **74[6]**, 1335-1339 (1991).

[11]T. M. Shaw and B. A. Pethica, "Preparation and Sintering of Homogeneous Silicon Nitride Green Compacts," *J. Am. Ceram. Soc.*, **69[2]**, 88-93 (1986).

[12]D. Sciti and A. Bellosi, "Effects of Additives on Densification, Microstructure and Properties of Liquid-Phase Sintered Silicon Carbide," *J. Mater. Sci.*, **35**, 3849-3855 (2000).

[13]G. D. Zhan, R. J. Xie, M. Mitomo and Y. W. Kim, "Effect of β-to-α Phase Transformation on the Microstructure Development and Mechanical Properties of Fine-Grained Silicon Carbide Ceramics," *J. Am. Ceram. Soc.*, **84[5]**, 945-950 (2001).

[14]J. H. She and K. Ueno, "Densification Behavior and Mechanical Properties of Pressureless-Sintered Silicon Carbide Ceramics with Alumina and Yttria Additions," *Mater. Chem. Phys.*, **59**, 139-142 (1999).

[15]G. Magnani, G. L. Minoccari and L. Pilotti, "Flexural Strength and Toughness of Liquid Phase Sintered Silicon Carbide," *Ceram. Inter.*, **26**, 495-500 (2000).

[16]V. D. Krstic, "Optimization of Mechanical Properties in SiC by Control of the Microstructure," *Mater. Res. Soc. Bull.*, **20**, 46-49 (1995).

[17]J. H. She and K. Ueno, "Effect of Additive Content on Liquid-Phase Sintering on Silicon Carbide Ceramics," *Mater. Res. Bull.*, **34**, 1629-1636 (1999).

[18]S. K. Lee and C. H. Kim, "Effect of α-SiC versus β-SiC Stating Powders on Microstructure and Fracture Toughness of SiC Sintered with Al_2O_3-Y_2O_3 Additives," *J.Am.Ceram.Soc.*, **77[6]**, 1655-1658 (1994).

[19]S. G. Lee, W. H. Shim, J. Y. Kim, Y. W. Kim and W. T. Kwon, "Effect of Sintering-Additive Composition on Fracture Toughness of Liquid-Phase-Sintering SiC Ceramics," *J. Mater. Sci. Lett.*, **20**, 143-146 (2001).

[20]K. S. Cho, H. J. Choi, J. G. Lee and Y. W. Kim, "R-Curve Behavior of Layered Silicon Carbide Ceramics with Surface Fine Microstructure," *J. Mater. Sci.*, **36**, 2189-2193 (2001).

[21]J. Y. Kim, H. G. An, Y. W. Kim and M. Mitomo, "R-curve Behavior and Microstructure of Liquid-Phase Sintered α-SiC," *J. Mater. Sci.*, **35**, 3693-3697 (2000).

[22]H. Ye, V. V. Pujar and N. P. Padture, "Coarsening in Liquid-Phase-Sintered α-SiC," *Acta Mater.*, **47[2]**, 481-487 (1999).

[23]Y. W. Kim, M. Mitomo and G. D. Zhan, "Mechanism of Grain Growth in Liquid-Phase-Sintered β-SiC," *J. Mater. Res.*, **14[11]**, 4291-4293(1999).

[24]S. Tabata and Y. Hirata, "Colloidal Processing of SiC with 700 MPa of Flexural Strength," *Ceram. Trans.*, **152**, 119-128 (2004).

[25]Y. Hirata and W. H. Shih, "Colloidal Processing of Two-Component Powder System"; pp.637-644 in Advances in Science and Technology 14, Proceedings of 9 th Cimtec-World Ceramics Congress, Ceramics : Getting into the 2000's-Part B, Edited by P. Vincenzini, Techna Srl., 1999.

[26]N. Hidaka, Y. Hirata and S. Sameshima, "Colloidal Processing of the SiC-Al_2O_3-Y^{3+} Ions System and Sintering Behavior of the Consolidated Powder Compacts," *J. Ceram. Proc. Res.*, **3**, 271-277 (2002).

[27]N. Hidaka and Y. Hirata, "Colloidal Processing and Liquid Phase Sintering of SiC-Al_2O_3-Y^{3+} Ions System," *Ceram. Trans.*, **152**, 109-118 (2004).

[28]N. Hidaka and Y. Hirata "Mixing Effect of Nanometer-Sized SiC Powder on Processing and Mechanical Properties of SiC Using Submicrometer-Sized Powder", *J. Ceram. Soc. Japan*, **113[7]**, 466-472 (2005).

[29]Japanese Industrial Standards R 1601, "Testing Method for Flexural Strength of Fine Ceramics", Japanese Industrial Standards Committee, 1995.

[30]Y. Hirata, S. Tabata and J. Ideue, "Interaction of the Silicon Carbide-Polyacrylic Acid-Yttium Ions System," *J. Am. Ceram. Soc.*, **86[1]**, 5-11 (2003).

[31]C. F. Baes Jr. and R. E. Mesuner, The Hydrolysis of Cations; p. 134. Robert E. Krieger Pub., Malabar, Florida, 1986.

[32]Y. Hirata, J. Kamikakimoto, A. Nishimoto and Y. Ishihara, "Interaction between α-Alumina Surface and Polyacrylic Acid," *J. Ceram. Soc. Japan*. **100[1]**, 7-12(1992).

[33]E. M. Levin, C. R. Robbins and H. F. McMurdie, Phase Diagrams for Ceramists, 1969 Supplement, p. 165. Edited by M. K. Reser. The American Ceramic Society, Columbus, OH, 1969.

[34]R. A. Swalin, "Thermodynamics of Solid," pp.148-152, pp. 180-184. John Wiley & Sons, New York, 1972.

MECHANICAL PROPERTIES OF Ce-DOPED ZIRCONIA CERAMICS SINTERED AT LOW TEMPERATURE

Michihito Muroi and Geoff Trotter
Advanced Nanotechnology Limited
112 Radium Street
Welshpool, WA 6106, Australia

ABSTRACT

Ce-doped zirconia nanopowders having an average particle size of about 20 nm have been synthesized by a technique based on mechanochemical processing, a process that makes use of chemical reaction effected by high-energy ball milling and subsequent heat treatment. It was found that the zirconia nanopowders could be sintered to near full density at temperatures below 1150°C within several hours even when the green bodies were prepared by uniaxial pressing at a low pressure of 50 MPa. The sintered ceramics had a relatively narrow grain-size distribution with the average grain size in the range 150-300 nm depending primarily on sintering temperature. Measurements of mechanical properties showed that for fully dense tetragonal ceramics the Vickers hardness increased somewhat with Ce concentration, while the fracture toughness depended strongly on the Ce concentration with a maximum of about 18 MPa·m$^{1/2}$ observed near the tetragonal-monoclinic phase boundary. The three-point bending strength was in the range 450-750 MPa for tetragonal ceramics, and nonlinear stress-strain curves were observed for the compositions corresponding to high fracture toughness. In view of the superior mechanical properties of sintered ceramics, as well as the lower pressure for powder consolidation and the lower sintering temperature, the results of this work are considered to be potentially an important advance towards the use of nanopowders for mass production of zirconia ceramics.

INTRODUCTION

Zirconia, especially in the form fully or partially stabilized by doping with stabilizers such as Y_2O_3, CeO_2, CaO and MgO, is one of the most important ceramic materials, being used in a wide range of applications owing to its unique properties. Fully stabilized zirconia, for instance, is used as an active material of oxygen sensors and an electrolyte of solid oxide fuel cells (SOFC), taking advantage of its high ionic conductivity. Unlike most other engineering ceramics, which are hard and strong but brittle, partially stabilized zirconia possesses high fracture toughness and wear resistance as well as high hardness and strength, and is widely used in demanding applications such as cutting tools, electronic components, engine components, grinding media and optical-connector parts.

In almost all cases zirconia ceramics are produced through sintering of green bodies, i.e. consolidated powders. Because of the refractory nature of zirconia, the sintering of conventional micron-sized powders requires a high temperature well over 1500°C. The use of sub-micron powders, now commercially available, allows sintering at lower temperatures, typically in the range 1400-1500°C, but a further reduction in sintering temperature is desirable to reduce production cost and to minimize grain growth during sintering.

One of the most effective ways of lowering sintering temperature is to reduce the particle size, thereby increasing the driving force for surface-area reduction. Studies have shown that the use of nanopowders enables sintering to near full density at considerably lower temperatures, down to around 1050°C, or even 950°C if sintering is carried out in a vacuum or inert atmosphere.[1-6] Nonetheless, nanopowders are not currently used for mass production of zirconia ceramics. The primary reason for this is that the preparation of homogeneous nanocrystalline green bodies, a prerequisite for low-temperature sintering, is difficult because of the tendency of nanoparticles to form hard agglomerates. In most previous studies this problem was overcome by a forcible means, viz. by consolidating the powder at exceedingly high pressures, ranging from around 500 MPa to 3 GPa.[2-6] However, this approach is impractical as it is applicable only to the preparation of very small samples of simple shape. Centrifugal consolidation was also successfully used for the preparation of homogeneous nanocrystalline green bodies that could be sintered to near full density at temperatures below 1100°C,[1] but this technique is also impractical because of low production rates and difficulty in automation.

In the present work, zirconia ceramics were prepared using nanopowders synthesized by a technique based on mechanochemical processing (MCP^{TM}),[7,8] which has proved to be effective in producing agglomerate-free nanopowders of a variety of materials, including transition metals,[9] oxide ceramics,[10-13] sulfide semiconductors,[14] and various kinds of magnetic materials.[15-17] It was found that the zirconia powders thus synthesized could be sintered to near full density at temperatures in the range 1100-1200°C within several hours even when the powders were consolidated at a low pressure of 50 MPa. The mechanical properties of ceramics sintered at low temperatures were investigated, and are reported in this paper.

EXPERIMENTAL

The zirconia nanopowders used in the experiment were synthesized by a technique based on MCP^{TM}. In this process, chemical precursors undergo reaction, during either milling or subsequent heat treatment, to form a nanocrystalline composite consisting of nanoparticles of the desired phase embedded in a salt matrix. The nanopowder is then recovered by removing the salt through a simple washing procedure. Details about the synthesis of nanopowders by MCP^{TM}, including that of zirconia nanopowder, are described in Refs. 7-17.

The powders thus synthesized had BET surface areas of about 50 m^2/g, corresponding to a particle size of about 20 nm. The powders were consolidated into green bodies by uniaxial pressing. For most experiments hardened-steel dies having a diameter of either 5 mm or 12.7 mm were used to prepare circular samples 2-3 mm in thickness. For bending-strength measurements, which require larger samples, a circular die having a diameter of 30 mm or a rectangular die having dimensions of 5 x 50 mm^2 was used. The green bodies were pressureless sintered in air at various temperatures; the heating and cooling rates were kept at 300°C/h. A dilatometer was used to study sintering behavior in depth.

The sintered samples were characterized for density (by the Archimedes method), crystal structure and phases present [by X-ray diffraction (XRD)] and microstructure [by scanning electron microscopy (SEM)].

The Vickers hardness and fracture toughness were measured using a standard indentation technique. A load of 50 kg was applied to a polished surface of a sintered pellet for 15 sec to make an indent. The Vickers hardness (H$_V$) was calculated on the basis of the equation $H_V = 1.854 P/a^2$, where P is the load and a is the diagonal length of the indent. The fracture toughness (K$_{IC}$) was derived using the equation $K_{IC} = 9.052 \times 10^{-3} \cdot H_V^{3/5} \cdot E^{2/5} \cdot a \cdot c^{-1/2}$, where E is Young's modulus assumed to be 200 GPa, a is the diagonal length of the indent, and c is the crack lengths.[18]

For three-point bending testing, bar-shaped samples having typical dimensions of 1.5 x 4 x 25 mm^3 were used with a support span of around 20 mm. A constant crosshead speed of 0.5

mm/min was employed. The bending strength (σ) was calculated using the standard equation $\sigma=3PF/(2WD^2)$, where F is the force at rupture, L the support span, W the sample width, and D the sample thickness.

RESULTS AND DISCUSSION

Powder consolidation and sintering behavior

In Fig. 1 the densities of 6%Ce samples before and after sintering (1125°C x 3 h) are plotted as a function of uniaxial pressure used for powder consolidation. (Here and in the following discussion, a sample containing x mol% CeO$_2$ is referred to as an "x%Ce sample".) The data for a 10 nm zirconia powder having the same composition, synthesized by a conventional coprecipitation technique, are also shown for comparison. It can be seen that for a given pressure the green density is considerably higher for the powder synthesized by MCP^{TM} ("MCP powder" hereafter) than for that synthesized by coprecipitation ("CP powder" hereafter); and that a green density of around 45%, generally considered necessary to achieve full densification, is attained even at 50 MPa for the MCP powder.

The difference in sintered density is still more remarkable. For the MCP powder the density is almost independent of pressure, and a near full density (97.5% assuming a theoretical density of ~6.1 g/cm^3) is achieved even for 50 MPa; the densities are essentially 100% for pressures greater than 100 MPa. For the CP powder, by contrast, the density increases considerably with increasing pressure and reaches only 82% even for a very high pressure of 1.4 GPa. It should be noted that it is not the higher green density itself that makes the MCP-powder compacts far more sinterable than the CP-powder compacts; in fact, the CP-powder compacts pressed at higher pressures have higher green densities than MCP-powder compacts pressed at lower pressures, and yet the sintered densities for the CP-powder are significantly lower than those for the MCP-powder. It is most likely that the high-density CP-powder compacts consist of agglomerates having a very high density and larger inter-agglomerate pores, while the MCP-powder compacts, even when pressed at a lower pressure, consist of uniformly distributed nanoparticles.

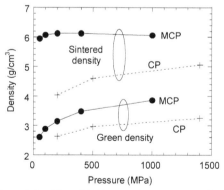

Fig. 1. Green and sintered densities of 6%Ce samples plotted as a function of uniaxial pressure applied for powder compaction. The green bodies were sintered at 1125°C for 3 h in air.

MCP: powder synthesized by mechanochemical processing. CP: powder synthesized by coprecipitation.

It has been reported in a number of publications that uniaxially pressed zirconia-nanopowder compacts can be sintered to near full density at low temperatures (<1200°C).[2-6] However, in these studies the powders were invariably pressed at a high pressure, typically 500 MPa or higher. (In one study an enormous pressure of 3 GPa was applied to prepare sinterable green bodies.[4]) The ability to sinter to full density at a low temperature when consolidated at a low pressure thus seems unique to the MCP powders. This is of technological significance, since mass production of larger ceramic articles by simple uniaxial die pressing is possible only if the pressure required for powder consolidation is not exceedingly high.

Figure 2 shows thermal expansion (TE) curves for MCP powders having various Ce concentrations and for a commercial $2.5mol\%Y_2O_3-ZrO_2$ (YSZ) powder; the samples used for the measurements were prepared by uniaxial pressing at 150 MPa. For the MCP powders, the major shrinkage occurs in a narrow temperature range between 1000 and 1150°C. The shrinkage shifts slightly towards higher temperature with increasing Ce concentration, but the densification is essentially completed during the 5 h holding at 1150°C for all three samples. (The sintered densities for the 3%Ce, 9%Ce and 12%Ce samples were 5.76, 6.08 and 6.10 g/cm^3, respectively. The lower density of the 3%Ce sample is due to a tetragonal-monoclinic phase transformation that occurred during cooling, as evidenced by the jump at about 570°C observed in the cooling section of the curve.) The TE curves for the commercial YSZ powder show that shrinkage occurs more gradually and in a higher temperature range than for the MCP powders. In the former sample, sintering was far from complete at 1150°C (the sintered density was only 3.85 g/cm^3), and a considerably higher temperature of 1400°C was required to achieve full densification. It should be noted that the MCP-powder compacts exhibit virtually no shrinkage up to about 800°C, but once densification starts it proceeds to completion in a relatively narrow temperature range. This again indicates that the constituent nanoparticles are uniformly packed in the green body in spite of the low pressure (150 MPa) used for powder consolidation. (If the particle distribution is not uniform, large pores will be formed in the intermediate stage of sintering, thus preventing full densification in the final stage;[1] if the particles are packed uniformly but loosely, particle rearrangement will occur in the initial stage of sintering, resulting in noticeable shrinkage in the lower temperature range.[2])

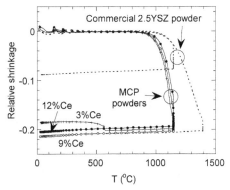

Fig. 2. Thermal expansion curves for MCP powders having various Ce concentrations and for a commercial 2.5mol%Y_2O_3-ZrO_2 powder. The pellets, prepared by uniaxial pressing at 150 MPa, were subjected to a thermal cycle consisting of heating (at 300°C/h), holding (at 1150 or 1400°C for 5 h) and cooling (at 300°C/h) segments.

The ability to consolidate MCP powders into a homogeneous green body at a relatively low pressure can be ascribed to the following three factors. First, MCP powders in general have a very low degree of particle agglomeration, a feature that has enabled commercialization of various MCP-nanopowder products, such as CeO_2 chemical-mechanical-polishing slurries and transparent ZnO UV-light absorbers. Second, the nanoparticles constituting the MCP powders mostly have quasi-spherical morphology, as confirmed by TEM. This, combined with the low degree of particle agglomeration, provides the MCP powders with better flowability, thus facilitating consolidation into homogeneous green bodies. The third factor is related to surface properties. In MCP^{TM} there is no grinding step after crystalline nanoparticles are formed.[*] The nanoparticles thus have annealed surfaces, which are expected to be chemically and thermodynamically stable. As a consequence, inter-particle interaction and friction, which tend to oppose particle-particle sliding and rearrangement during die filling and pressing, will be relatively weak. On the other hand, in the synthesis of nanopowders by most other methods, including coprecipitation and hydrolysis, the crystalline zirconia particles formed through thermal decomposition of precursor materials are usually subjected to some form of ball milling or bead milling; this is necessary to break down agglomerates and/or aggregates formed during the preceding calcination step. Thus, the surfaces of the nanoparticles are mostly fracture surfaces, which are expected to be chemically and thermodynamically less stable than annealed surfaces. The resultant stronger inter-particle interaction and friction, as well as the more irregular particle morphology resulting from fracture, will make particle-particle sliding and rearrangement during die filling and powder compaction more difficult. Nanopowder synthesis from a gas phase, e.g. inert gas condensation[2] and chemical vapor synthesis,[3] do not necessarily involve a milling step. In these methods, however, crystalline particles are formed under nonequilibrium conditions, i.e. through rapid condensation from a gas phase. The properties of the surfaces of these particles are therefore expected to be similar to those of fracture surfaces.

Microstructure and crystal structure of sintered ceramics

Figure 3 shows SEM images of fracture surfaces of (a) a 6%Ce sample sintered at 1125°C for 3 h and (b) a 9%Ce sample sintered at 1100°C for 8 h. It can be seen that both samples are fully dense and have uniform, fine-grained microstructure. The average grain size of the 9%Ce sample (~300 nm) is greater than that of the 6%Ce sample (~150 nm), reflecting the somewhat higher sintering temperature for the former.

Fig. 3. SEM images of fracture surfaces of (a) a 6%Ce sample sintered at 1125°C for 3 h (ρ=6.12 g/cm^3) and (b) a 9%Ce sample sintered at 1180°C for 8 h (ρ=6.21 g/cm^3).

XRD measurements showed that sintered 3%Ce samples were predominantly monoclinic zirconia, while the sintered 9%Ce and 12%Ce samples were essentially 100% tetragonal zirconia, regardless of sintering temperature up to 1200°C. The crystal structure of 6%Ce samples, however, depended on the sintering temperature. The samples sintered at temperatures up to 1150°C were essentially 100% tetragonal, while the sample sintered at 1200°C was predominantly monoclinic. These observations suggest that the critical Ce concentration corresponding to the tetragonal-monoclinic phase boundary is about 6mol%, a value lower than that for coarse-grained Ce-doped zirconia ceramics, 10-12mol%. The lower critical concentration in the present system is ascribed to the smaller grain size, which increases the stability of tetragonal phase relative to that of monoclinic phase. A similar reduction in the critical stabilizer concentration with decreasing grain size has also been observed in the Y-ZrO$_2$ system and explained in terms of a model in which an interfacial energy is included in the total free energy.[6]

Mechanical properties

Figure 4 shows the Ce-concentration dependence of the Vickers hardness (H$_V$) and fracture toughness (K$_{IC}$) of Ce-ZrO$_2$ samples. The samples were sintered at 1125°C for 3 h, except for the 12%Ce samples, which were sintered at 1150°C for 5 h to achieve full density. (A 12%Ce sample sintered at 1125°C had a lower density of 5.96 g/cm^3 and a lower H$_V$ of 9.0 GPa.) The 3%Ce samples were about 95% monoclinic, while the 6%Ce, 9%Ce and 12%Ce samples were fully tetragonal.

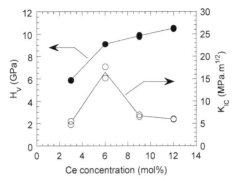

Fig. 4. Vickers hardness (H_V) and fracture toughness (K_{IC}) plotted as a function of Ce concentration. The samples were sintered at 1125°C for 3 h except for 12%Ce samples, which were sintered at 1150°C for 5 h to achieve full density.

For Ce concentrations higher than 6mol%, where the samples are fully tetragonal, H_V increases slightly with Ce concentration, from 9.1 GPa for 6mol% to 10.5 GPa for 12mol%. As will be discussed later, this Ce-concentration dependence of H_V is likely to be a result of decreasing transformability with increasing Ce concentration, rather than a direct effect of changing dopant concentration. The lower H_V (5.9 GPa) for the 3%Ce sample is ascribed to the fact that it consists almost entirely of monoclinic zirconia formed through a tetragonal-monoclinic phase transformation during cooling (Fig. 2); thus the sample is most likely to be microcracked, although it remained in one piece and had a near full density (5.76 g/cm³).

The K_{IC} peaks at a Ce concentration of 6mol% (15.5-17.7 MPa·m$^{1/2}$) and decreases with either decreasing or increasing Ce concentration. The high K_{IC} for the 6%Ce sample is due to the transformation toughening effect, which is known to be most effective in tetragonal zirconia ceramics that are on the verge of spontaneous tetragonal-monoclinic phase transformation and hence have high transformability.[5] The high transformability of the 6%Ce sample is demonstrated by the following experimental observations:

(1) The stress-strain curve for a 6%Ce sample, recorded during the bending test and presented in Fig. 5, is nonlinear, involving a change in slope on reaching a critical stress. This clearly demonstrates plastic deformation associated with the stress-induced tetragonal-monoclinic phase transformation at higher loads. By contrast, the stress-strain curve for a 12%Ce sample is linear all the way to rupture.

(2) Grinding the surface of a 6%Ce sample with 80-grit sand paper caused part of the tetragonal phase to transform into a monoclinic phase, as can be seen clearly in the XRD patterns presented in Fig. 6.

(3) A transformed zone, similar to the one reported in Ref. 5, was observed around Vickers indentations.

The much lower K_{IC} for the 3%Ce sample is attributed to the fact that it is predominantly monoclinic zirconia, which has no ability to transform to a lower-density phase. The decrease in K_{IC} with increasing Ce concentration beyond 6mol%, on the other hand, is consequent upon progressive stabilization of the tetragonal phase and the resultant decrease in transformability. A similar Ce-concentration dependence of K_{IC} in the tetragonal field has been observed in a previous study and shown to have a good correlation with the amount of monoclinic phase formed through stress-induced transformation.[10]

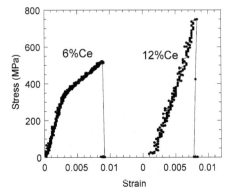

Fig. 5. Stress vs strain curves for a 6%Ce samples (sintered at 1125°C x 3 h) and a 12%Ce sample (sintered at 1150°C x 5 h). Strain=6yD/L², where y is the deflection, D the sample thickness and L the support span.

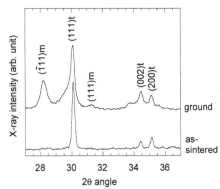

Fig. 6. XRD patterns for 6%Ce samples (sintered at 1150°C for 2 h) before and after surface grinding with 80-grit sand paper.

Measurements of bending strength (σ) showed that for essentially fully dense, tetragonal samples, σ decreased with decreasing Ce concentration as exemplified by the data in Fig. 5.[**] The lower bending strength of the 6%Ce sample can be ascribed to the high transformability (i.e. low critical stress needed to induce phase transformation), which leads to the formation of large zones of transformation under stresses during the bending test, causing damage to microstructure. A similar reduction in σ has been observed in Ce-doped zirconia ceramics that exhibit plastic deformation, and explained in terms of this model.[20] Transformation plasticity makes the ceramic flaw-tolerant, but seems to be detrimental to ultimate strength.[20]

Figure 7 shows the K_{IC} vs H_V relationship for a large number of predominantly tetragonal, essentially fully dense zirconia ceramics, including the Ce-doped samples the properties of which have been described above. The ceramics, having a wide variety of compositions, were prepared by sintering MCP nanopowders at temperatures between 1100 and 1200°C. The data clearly show that K_{IC} and H_V are in a trade-off relationship: a ceramic having a higher K_{IC} tends to have a lower H_V. This, combined with the high transformability of the 6%Ce samples as demonstrated above, suggests that the plastic deformation brought about by stress-induced phase transformation is responsible for the decrease in H_V with increasing K_{IC}.

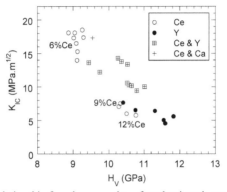

Fig. 7. K_{IC} vs H_V relationship for a large number of predominantly tetragonal, essentially fully dense zirconia ceramics having various types of stabilizers. All the samples were prepared by uniaxially pressing MCP nanopowders at 100-200 MPa, followed by sintering at 1100-1200°C.

It should be noted that while a good correlation between K_{IC} and H_V is observed in Fig. 7, it is certainly not described by a single curve. Particularly noteworthy is the fact that the ceramics doped with Y or Ca together with Ce tend to have superior mechanical properties. For instance, the modest values of K_{IC} of around 6-7 MPa·m$^{1/2}$ for the Ce-doped ceramics having an H_V of around 10.5 GPa are increased significantly (to 10-14 MPa·m$^{1/2}$) by co-doping with Y or Ca. Thus, a ceramic having improved mechanical properties can be developed through materials designing within the framework of the present technology.

CONCLUSIONS

Ce-doped zirconia ceramics have been prepared by the simple uniaxial pressing of a dry powder followed by pressureless sintering in air, using nanopowders synthesized by a technique based on mechanochemical processing (MCP^{TM}). It was found that the nanopowders thus synthesized could be sintered to near full density at temperatures below 1150°C even when they were consolidated into green bodies at a low pressures of 50 MPa, which is an order of magnitude lower than the pressure required to prepare sinterable green bodies from zirconia nanopowders synthesized by other methods. This indicates that particle-particle sliding and rearrangement during pressing are much easier for the MCP powders, allowing the formation of a homogeneous green body at a relatively low pressure. The ease of compaction of the MCP powders into a sinterable green body is ascribed tentatively to the low degree of particle agglomeration and to the uniformity and stability of the particle surfaces.

Measurements of mechanical properties showed that for essentially fully dense tetragonal ceramics the Vickers hardness (H_V) increased slightly with Ce concentration, while the fracture toughness (K_{IC}) depended strongly on Ce concentration with a maximum of about 18 MPa·m$^{1/2}$ observed for 6mol%, which is close to the tetragonal-monoclinic phase boundary. The Ce-concentration dependence of H_V, as well as that of K_{IC}, is argued to be related to stress-induced tetragonal-monoclinic phase transformation. The three-point bending strength (σ) increased with Ce concentration and a maximum value of 750 MPa was recorded for 12mol%. The Ce-concentration dependence of σ, which is qualitatively similar to that of H_V, can also be explained in terms of stress-induced phase transformation.

In view of the superior mechanical properties of sintered ceramics, as well as the lower pressure for powder consolidation and the low sintering temperature, the technology developed in this work could open up the possibility of using nanopowders in mass production of zirconia ceramics.

FOOTNOTES
*No grinding step is necessary in MCP^{TM}, since the nanoparticles formed are separated by a salt matrix. The only remaining step after heat treatment is a washing procedure to remove the salt.

**The bending strength of the 3%Ce samples, which were predominantly monoclinic, was not only lower than that of tetragonal samples, but also exhibited considerable scattering. This is most likely to be due to microcracks formed through the phase transformation during cooling (Fig. 2).

REFERENCES
[1] W. H. Rhodes, "Agglomerate and Particle Size Effects on Sintering Yttria-Stabilized Zirconia", *J. Am. Ceram. Soc.*, **64**, 19-22 (1981).
[2] G. Skandan, "Processing of Nanostructured Zirconia Ceramics", *NanoStruct. Mater.* **5**, 111-126 (1995).
[3] V. V. Sardic, M. Winterer, and H. Hahn, "Sintering Behavior of Nanocrystalline Zirconia Prepared by Chemical Vapor Synthesis", *J. Am. Ceram. Soc.*, **83**, 729-736 (2000).
[4] L. Gao, W. Li, H. Z. Wang, J. X. Zhou, Z. J. Chao, and Q. Z. Zai, "Fabrication of Nano Y–TZP Materials by Superhigh Pressure Compaction", *J. Eur. Ceram. Soc.* **83**, 135-138 (2001).
[5] A. Bravo-Leon, Y. Morikawa, M. Kawahara, and M. J. Mayo, "Fracture Toughness of Nanocrystalline Tetragonal Zirconia with Low Yttria Content", *Acta Materialia* **50**, 4555-4562 (2002).
[6] M. J. Mayo, A. Suresh, and W. D. Porter, "Thermodynamics for Nanosystems: Grain and Particle-Size Dependent Phase Diagrams", *Rev. Adv. Mater. Sci.*, **5**, 100-109 (2003).
[7] P. G. McCormick, J. Ding, W-F. Miao, and R. Street, "Process for the Production of Ultrafine Particles", United States Patent 6203768 B1 (2001).
[8] P. G. McCormick, P. Gerard, and T. Tsuzuki, "Process for the Production of Ultrafine Powders of Metal Oxide", United States Patent 6503475 (2003).

[9]J. Ding, T. Tsuzuki, P. G. McCormick, and R. Street, "Ultrafine Co and Ni Particles Prepared by Mechanochemical Processing", *J. Phys. D - Applied Physics*, **29**, 2365-2369 (1996).

[10]J. Ding, T. Tsuzuki, and P. G. McCormick, "Ultrafine Alumina Particles Prepared by Mechanochemical/Thermal Processing", *J. Am. Ceram. Soc.*, **79**, 2956-2958 (1996).

[11]A. C. Dodd, K. Raviparasad, and P. G. McCormick, "Synthesis of Ultrafine Zirconia Powders by Mechanochemical Processing", *Scripta Mater.* **44**, 689-694 (2001).

[12]A. C. Dodd and P. G. McCormick, "Solid-State Chemical Synthesis of Nanoparticulate Zirconia", Acta Mater. **49**, 4215-4220 (2001).

[13]A. C. Dodd and P. G. McCormick, "Synthesis of Nanocrystalline ZrO_2 Powders by Mechanochemical Reaction of $ZrCl_4$ with LiOH", *J. Eur. Ceram. Soc.* **22**, 1823-1829 (2002).

[14]T. Tsuzuki and P. G. McCormick, "Synthesis of CdS Quantum Dots by Mechanochemical Reaction", *Appl. Phys. A* **65**, 607-609 (1997).

[15]W. Liu and P. G. McCormick, "Formation and Magnetic Properties of Nanosized Sm_2Co_{17} Magnetic Particles via Mechanochemical/Thermal Processing", *NanoStruct. Mater.* **12**, 187-190 (1999).

[16]M. Muroi, R. Street, and P. G. McCormick, "Enhancement of Critical Temperature in Fine $La_{0.7}Ca_{0.3}MnO_3$ Particles Prepared by Mechanochemical Processing", *J. Appl. Phys.* **87**, 3424-3431 (2000).

[17]M. Muroi, R. Street, P. G. McCormick, and J. Amighian, "Magnetic Properties of Ultrafine $MnFe_2O_4$ Powders Prepared by Mechanochemical Processing", *Phys. Rev. B* **63**, 184414-1-7 (2001).

[18]K. E. Amin, "Toughness, Hardness, and Wear", in *Ceramics and Glasses,* Engineering Materials Handbook Vol. 4, ASM International, pp601 (1991).

[19]K. Tsukuma and M. Shimada, "Strength, Fracture Toughness and Vickers Hardness of CeO_2-stabilized Tetragonal ZrO_2 polycrystals (Ce-TZP)", *J. Mater. Sci.* **20**, 1178-1184 (1985).

[20]G. Grathwohl and T. Liu, "Crack Resistance and Fatigue of Transforming Ceramics: II, CeO_2-Stabilized Tetragonal ZrO_2", *J. Am. Ceram. Soc.*, **74**, 3028-3034 (1991).

USING MASTER CURVE MODEL ON THE SINTERING OF NANOCRYSTALLINE TITANIA

Mao-Hua Teng* and Mong-Hsia Chen
Department of Geosciences, National Taiwan University,
No.1, Sec.4, Roosevelt Road
Taipei, Taiwan 106, R.O.C.

ABSTRACT

Due to size effects, the sintering behavior of nanocrystalline ceramic powders is much more complicated than that of a micron-sized ceramic powder. Though some may disagree, to date, no practical model has been developed to describe the sintering of nanoceramics. The authors have previously developed a Master Curve Model, derived from general chemical kinetic reactions, and have successfully applied the model to predict the sintering behavior of both micron and submicron powders. Recently, we've conducted a feasibility study using the Master Curve Model to describe and predict the sintering behavior of nanocrystalline titania powder. Considering many complicated factors, such as the fast surface diffusion, severe agglomeration, a phase transformation, and other nano-size effects, it seems that the Master Curve Model will be impossible to describe the sintering of nanocrystalline ceramic powders. To our surprise, the master curve relationship clearly exists for the powder that has gone through the phase transformation; however, the values of the apparent activation energy derived from the model are higher than that one would expected from conventional grain boundary and volume diffusion. Furthermore, the apparent activation energy seems inversely proportional to the logarithmic size of the titania particles. We believe this relationship may provide valuable clues to the sintering mechanisms, but have yet to find a satisfactory explanation for it.

INTRODUCTION

Though many conventional sintering models have been developed to predict and describe microstructure evolution during the sintering of ceramic powder compacts, most models are specifically applicable only to initial-, intermediate-, or final-stage of sintering.[1] Additionally, the parameters used in these models are usually very hard to measure, making their applications impractical [1, 2]. In 1996, Su and Johnson [3] developed a more practical Master Sintering Curve Model (MSC) that can be used to describe the sintering of many micron-sized ceramic powders. However, for nanocrystalline ceramic powders, there is still no practical and effective sintering model. This is possibly due to a more complicated combination of sintering mechanisms in nano-size ceramics as compared to micron size ceramics [4, 5].

Recently we've derived a mathematical model from the general chemical kinetics equation. The final mathematic equation looks very similar to the MSC Model [3], i.e. the MSC has an additional 1/T term in the integration part of its equation, and therefore, we named it after MSC and called it Master Curve Model (MCM.) The model can be applied universally in the analysis of many kinetic reactions -- including sintering. Though the Master Curve Model has proved to be very effective in the analysis of the sintering of micron-sized and submicron-sized ceramics [6], we still don't know weather it can be applied to nanocrystalline ceramics. Considering many complicated factors, such as the fast surface diffusion, severe agglomeration, and other nano-size effects, at first it would seem that it would be impossible for the Master Curve Model to be able to describe the sintering of nanocrystalline ceramic powders; however, our preliminary results showed that the model can indeed be applied to the sintering of nanocrystalline α-Al₂O₃.[7]

The purpose of this study is to test the applicability of the MCM to the sintering of another nanoceramic powder--nanocrystalline titania. During sintering, the nano-sized anatase-phase titania particles will transform into stable rutile phase titania. With this additional complication, i.e. a phase transformation, will the model still be applicable? In the following sections we will first give a brief introduction to the MCM.

MASTER CURVE MODEL (MCM)

The Master Curve Model [6] was derived from the general chemical kinetics equation:

$$\frac{d\alpha}{dt} = k(T)f(\alpha) \tag{1}$$

where α is the fraction of reactant (but it could also be relative density in the case of sintering as in this study), $f(\alpha)$ is dependant on the reaction mechanism, e.g., $f(\alpha) = (1-\alpha)$ when the reaction is first order, and $f(\alpha) = (1-\alpha)^2$ when the reaction is second order, and so on. t is the reaction time, $k(T)$ is the Arrhenius constant, which usually has the form $k(T) = A \cdot \exp(-Q_a/RT)$, where A is the collision factor, T is the absolute temperature, R is the gas constant, and Q_a is the apparent activation energy. Rearranging Eq.(1) --

$$\frac{d\alpha}{f(\alpha)} = k(T)dt = A \cdot \exp\left(\frac{-Q_a}{RT}\right)dt \tag{2}$$

and integrating both sides of Eq.(2), we have:

$$\int_0^\alpha [f(\alpha)]^{-1}d\alpha = A \int_0^t \exp\left(\frac{-Q_a}{RT}\right)dt \tag{3}$$

Let the right side of Eq.(3), which related to temperature and time, be Σ:

$$\Sigma(t, T(t)) \equiv \int \exp\left(\frac{-Q_a}{RT}\right)dt \tag{4}$$

and let the left side, which related to the fraction of reaction, be Φ:

$$\Phi(\alpha) = \frac{1}{A}\int_0^\alpha [f(\alpha)]^{-1}d\alpha \tag{5}$$

If we calculate Σ by integrating the temperature and time of the reaction, and make a plot with $\log(\Sigma)$ as abscissa and α as ordinate, all of the data will fall on an S-shaped curve, i.e. the master curve. The critical part of this calculation is to determine the best Q_a for the experimental data that gives the best fit to the master S curve. Comparing Eq.(4) to the equation for the Mastering Sintering Curve by Su and Johnson [3]:

$$\Theta(t, T(t)) \equiv \int_0^t \frac{1}{T} \exp\left(\frac{-Q_a}{RT}\right) dt \qquad (6)$$

the two models are almost the same, except for the additional $1/T$ term in Eq.(6). Note that Eq.(6) was derived from the combined-stage sintering model [2], and therefore inherited several important assumptions; while Eq.(4) came from the general kinetic equation and has nothing to do with conventional sintering models. Nevertheless, we'll going to demonstrate in this paper that the MCM can indeed be used to analyze sintering experiments, although there is this $1/T$ difference between the two models.

EXPERIMENTAL

To test the applicability of Master Curve Model, we've done a series of sintering experiments on several different nanocrystalline ceramic materials. However, in this study we'll concentrate on the results of one anatase phase nanocrystalline titania powder.

Sample preparation

The anatase phase, >99% pure nanocrystalline TiO_2 powder (CHT-TiO-n01 used has a nominal average diameter of about 10 nm. Our own analysis showed that the BET specific area is 320 m^2/g ⌐ NOVA 2000, Quantachrom Co.⌐which corresponds to an equivalent average diameter 10 nm. X-ray diffraction analysis confirmed that the powder is anatase phase. After dispersing the as-received powder in methanol, ultrasonicating for two hours, and adding 1.2 wt.% PVA as binder, the powder was dried, sieved, and compacted following the usual powder preparation procedures. Sixty green compacts were formed (by uniaxial pressing) at 7 MPa to produce an initial relative density about 21.5%.

Sintering and density measurement

All the sintering experiments were carried out in a 1700°C tube furnace (LINDBERG/BLUE M). We divided the sixty compacts into three sets, each with a different heating path (3°C /min, 5°C /min, and 10°C /min) and the sampling conditions are shown in Figure 1. Starting from 900°C, samples were collected every 50°C. The density of the samples was measured using Archimedes method, and the values were then divided by the theoretical density 3.893 g/cm^3 of titania (anatase phase) to determine the relative density. Note that we didn't use rutile's density as the base to calculate relative density, even though all the initial anatase samples transform into rutile phase in the later stage of sintering.

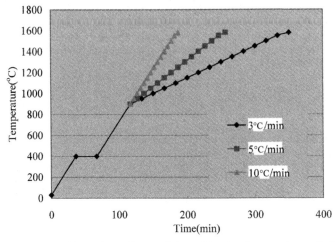

Figure 1. The heating profiles and sampling conditions of the sintering experiments in this study.

Data analysis

The data from the sintering experiments, including temperature, time, and relative density, were analyzed using a custom Visual Basic for Application (VBA) computer program tool. [8] The computer program automatically finds the best apparent activation energy and establishes the Master Curve relationship.

The original nano-size particles continuously grow during sintering. However, for the purpose of the MCM analysis, as long as all the green compacts experience the same path of geometric change, the data should be able to merge into one unique S-shape curve.

DISCUSSION

Master Curve analysis results

Figure 2 shows the sintering results for the nanocrystalline titania samples processed using the three heating paths shown in Figure 1. From the results, we clearly find a two-stage densification process. In the first stage, the samples start to densify as soon as they reach 800°C, and densification slows down at ~48% relative density. In the second stage, after reaching about 1250°C, the samples begin to densify again, until reaching their final densities above 97% (~90% for rutile's density) at 1550°C. X-ray diffraction analysis shows that the phase transformation is completed in the end of the first stage, i.e., the anatase has all been transformed into rutile.

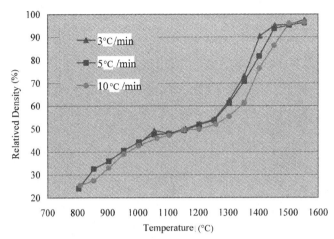

Figure 2. The change in relative density with temperature for the three different heating schedules shown in the Figure 1.

Because the MCM assumes a single S-shape curve relationship, the two-stage densification results in Figure 2 cannot be automatically merged into a single S curve by the Master Curve analysis [8] (Figure 3). However, if we separate the data into two groups: below and above 1050°C, representing the density before and after the phase transformation, and do the MCM analysis on each part we can easily derive an S-shape curve from the data above 1050°C (Figure 4); however not for the data below 1050°C. The equation of the S-shape master curve as shown in Figure 4 can be expressed as:

$$y = 47.74 + \frac{48.84}{[1 + \exp(-\frac{x - 13.24}{0.23})]^{0.26}} \tag{7}$$

The best apparent activation energy (Q_a), i.e., the one that minimizes the residual sum of square value, was found to be 519.5 kJ/mol. (Figure 5) This value, however, is much higher than values derived from isothermal sintering of submicron-size powders [9], and it is higher than the activation energy for grain-boundary diffusion or volume diffusion. This suggests that the apparent activation energy from the Master Curve Model is not one that has meaning that is consistent with conventional understanding.

The question then becomes why the activation energy is so high? It could reflect the effect of the small particle size, as shown in Figure 6. When we plot apparent activation energy vs. the log of titania particle size, we find a very nice inverse linear relationship. The titania powder in this study (i.e., 10 mm in diameter, and Qa–319.3 kJ/mol) is plotted at the upper left part of Figure 6. The other three larger size titania powders, which will not be discussed in this paper, have a relatively lower activation energy.

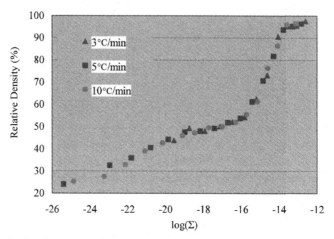

Figure 3. The sintering data for the nanocrystalline titania do not merge into a single S-shape Master Curve. Note that the data points less than 48% relative density are not randomly distributed.

But the linear relationship could also reflect the low initial density of the green samples. We are all familiar with the concept of sintering – to remove all the pores inside a green compact. Because in a green compact with about 20% relative density (such as in this study) all the particles are loosely packed, the atoms need to diffuse longer distance to fill up the pore. In addition, the surface diffusion becomes more important in the sintering of smaller particle, but it will not densify the sample. Other mechanisms, such as particle rearrangement (sliding, rotating…) could also be a factor when porosity is large. Due to the fact that nanocrystalline ceramic powders are very hard to press to high density, it may be the large pore volume that contributes to the abnormal high value of the apparent activation energy.

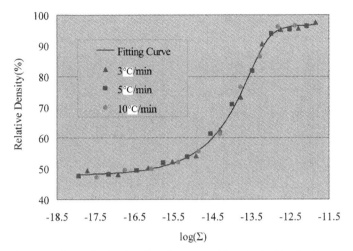

Figure 4. The Master Curve for the sintering of nanocrystalline titania after the phase transformation.

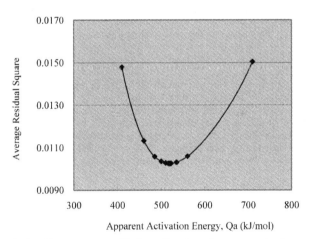

Figure 5. The calculated least squares residual as a function of apparent activation energy showing the minimum (i.e., best Qa) of 519.5 kJ/mol.

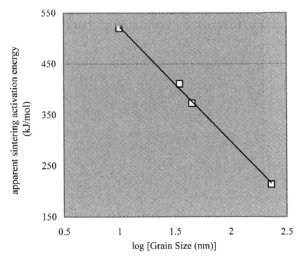

Figure 6. Apparent activation energy as a function of log particle size of titania powder.

Phase transformation results

The results indicate that the Master Curve Model cannot be applied to the analysis of the sintering of a low density nanocrystalline anatase-phase titania powder, i.e., at lower than 1050°C. The VBA computer program can derive an equation for a best-fit S-curve from the experimental data, however, the solution is not acceptable since the data points are not randomly distributed along the curve.

The question then becomes why does the MCM analysis of the phase transformation fail? The change in density during the first stage is significant (see Figure 2., from 20% to about 50% relative density). By comparison, the density difference between rutile and anatase is only about 8%. Therefore, the larger change in density cannot be explained just by the phase transformation alone. Thus, there must be densification due to sintering at the same time. Though MCM is a universal kinetic model, and it can analyze both the phase transformation and the sintering, it cannot analyze a single system with two different kinetic reactions going at the same time.

CONCLUSIONS

In addition to micron and submicronsized ceramic powders, we found that the Master Curve Model (MCM) can also be used to analyze the sintering of nanocrystalline titania. However, one cannot use MCM to analyze the sintering behavior of nanocrystalline titania below 1050°C when both a phase transformation and sintering occur simultaneously. Once the phase transformation is completed above 1050°C, one can easily derive an S-shape master curve relationship from the experimental data. We also found that the value of the derived apparent activation energy is much higher than the activation energy one would expect for conventional diffusion. This may indicate that nanocrystalline materials have more complicated sintering mechanisms than what we're familiar with a more conventional ceramic material; or it may simply reflect the consequences of the high initial porosity of the nanocrystalline titania green compacts.

ACKNOWLEDGMENTS

This work was supported by National Science Council, Taiwan, under the grant no. NSC93-2116-M-002-037.

REFERENCES

[1]R.L. Coble, "Sintering Crystalline Solids. I. Intermediate and Final Stage Diffusion Models," *Journal of Applied Physics*, **32**, 787-92 (1961).

[2]J.D. Hansen, R.P. Rusin, M.H. Teng and D.L. Johnson, "Combined-Stage Sintering Model," *J. Am. Ceram. Soc.*, **75**, 1129-35 (1992).

[3]H. Su and D.L. Johnson, "Master Sintering Curve: A Practical Approach to Sintering," *J. Am. Ceram. Soc.*, **79**, 3211-17 (1996).

[4]M.J. Mayo, "Processing of Nanocrystalline Ceramics from Ultrafine Particles," *International Materials Reviews*, **41**, 85-115 (1996).

[5]J.R. Groza, "Nanosintering," *NanoStructured Materials*, **12**, 987–992 (1999).

[6]C.H. Liang, "Three New Analytic Methods for Kinetic Reactions and Sintering," M.S. Thesis, National Taiwan University, 70-72, 2003 (in Chinese).

[7]M.H. Teng and M.H. Chen, "Using Master Curve Model on the Sintering of Nanocrystalline α-Al$_2$O$_3$," *IUMRS International Conference in Asia*, 306 (2004).

[8]M.H. Teng, Y.C. Lai and Y.T. Chen, "A Computer Program of Master Sintering Curve Model to Accurately Predict Sintering Results," *Western Pacific Earth Sciences*, **2**, 171-80 (2002).

[9]R.M. German, Sintering Theory and Practice, (John Wiley & Sons, New York, 1996) p.525.

MECHANICAL PROPERTIES AND HARDNESS OF ADVANCED SUPERHARD NANOCRYSTALLINE FILMS AND NANOMATERIALS

Murli H. Manghnani, Pavel V. Zinin, Sergey N. Tkachev
School of Ocean and Earth Science and Technology,
University of Hawaii
Honolulu, HI 96822, USA

Pavla Karvankova, Stan Veprek
Institute for Chemistry of Inorganic Materials,
Technical University Munchen
D-85747 Garching, Germany

ABSTRACT

The mechanical properties of materials such as hardness begin to deviate from bulk scaling laws when characteristic dimensions become extremely small. It was demonstrated recently that hardness of some nanocomposites films (nc-TiN/a-Si$_3$N$_4$ and nc-TiN/a-BN) were close to that of diamond. Here, we examine the relationships between the elastic properties, nano-structure, and hardness of advanced superhard nanocrystalline films *and* nanomaterials. We have investigated the elastic properties of the new cubic BC$_2$N phase and two types of nanocomposite films (nc-TiN/a-Si$_3$N$_4$ and nc-TiN/a-BN) by surface Brillouin scattering (SBS) in order to determine the elastic properties and quantify elastic moduli. It is shown that the values of the elastic moduli obtained from indentation measurements may significantly deviate from those measured by SBS.

INTRODUCTION

Many physical phenomena in both organic and inorganic materials have natural length scales between 1 and 100 nm (10^2 to 10^7 atoms). In many instances, the materials properties can be tuned by controlling the physical size of constituent grains. Like many other properties, the mechanical properties of materials begin to deviate from bulk scaling laws when characteristic dimensions become extremely small (nm size). Indeed, the mechanical properties of hard nanomaterials, especially hardness, could differ widely from their bulk properties. New nanocrystalline materials recently synthesized under high pressure and temperature exhibit extremely high hardness. The new BC$_2$N phase was found to be the second hardest material after diamond [1]. Popov *et al* [2] obtained a new superhard phase composed of single-wall carbon nanotubes (SP-SWNT) with a bulk modulus exceeding or comparable to that of diamond, and with hardness between those of cubic BN and diamond. Veprek demonstrated that hardness of the nanocomposites might be close to diamond [3].

Nanocomposite films are usually several microns thick. In such a case, only one elastic modulus (Young's modulus) can be determined from laser-SAW (surface acoustic waves) measurements, or acoustic microscopy [4] or vibrating-reed technique [5]. Surface Brillouin scattering (SBS) has proven to be a very effective method for measuring elastic properties of submicron films [6]. Usually, SBS allows measurements of the velocity of the Rayleigh surface acoustic wave (RW), and therefore estimation of the shear modulus of the material. However, in some cases, another wave (longitudinal lateral mode, or LM [6]) can be detected. Detection of the lateral waves provides an opportunity to obtain the complete set of the elastic moduli of isotropic material [7]. The objective of this paper is to study the elastic properties of advanced superhard nanocrystalline films and superhard nanomaterials using SBS, to compare the results with those obtained by nanoindentation, and to investigate the relationship between the elastic properties and hardness of these materials and films.

RESULTS

Nanofilms

The hardness of nanocomposite (nc) films (nc-TiN/a-Si$_3$N$_4$ and nc-TiN/a-BN) developed recently appears to approach that of diamond [8,9]. Moreover, the new nanocomposites show unusually large elastic moduli[9]. For example, Young's modulus of the Ti-B-N nanocomposite films measured by means of indentation technique is found [9] to be 607 GPa, which is higher than that of polycrystalline TiN (418 GPa[10]) and less than that of cubic boron nitride (cBN), 847 GPa[11]. It is natural to assume that the high hardness is due to granular structure of nanocomposites (Hall-Petch effect, discussed in detail under the DISCUSSION section). However, the range of grain sizes in nanocomposites is small (5-3 nm) and falls in the region where Hall-Petch effect does not seem to be valid [12]. As suggested in Ref. [9], the high value of the Young's modulus measured by indentation technique has been attributed to the extremely high pressure under the tip of the indenter. The explanation proposed was based [9] on the fact that the pressure under the tip of the indenter could reach an extremely high value (up to 10 GPa) [13]. It is also well known that the elastic moduli increase under high-pressure [14]. Thus the elastic moduli determined by the indentation technique seem be ambiguously higher than the realistic values.

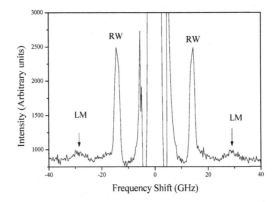

Figure 1. Experimental SBS spectrum (θ =45°) of nc-TiN/a-BN (HF110602) film; longitudinal lateral modes (LM) are shown by the arrows.

The indentation technique has been widely used for characterization of the mechanical properties of the newly developed hard and superhard films [15], and several new superhard phases. It is of fundamental importance to understand the physical phenomena operating during indentation in order to realistically estimate the relevant corrections necessary in the hardness measurement technique, particularly, in application to superhard materials.

The experimental Brillouin scattering setup we used has been adjusted to detect surface acoustic waves (SAW) and is an improved version of the one utilized in recent studies of bulk amorphous carbon samples[7]. A fully automated, self-aligning spectrometer with increased stability and flexibility consists of a Sandercock-type tandem six-pass Fabry-Perot interferometer. Beam from an argon ion laser (λ=514.5 nm and beam power of 130 mW) was focused on the film with 1:2.2 lens (f=50 mm). The same lens collects the scattered light in a backscattering geometry. A single-photon counting module converts detected light into series of electric pulses by a silicon avalanche photodiode. The Brillouin peak frequencies were determined by a curve-fitting routine. The frequency shift Δv from the SBS spectra is related to the SAW velocity V_{SAW} as

$$V_{SAW} = \frac{\lambda \Delta v}{2 \sin \theta} \qquad (1)$$

where θ is the scattering angle. Figure 1 shows SBS spectra obtained from three nc-TiN/a-BN films (HF110602); curve fitting was applied to extract positions of the LM peaks. These measurements provide bulk longitudinal velocities independently from refractive index [6,7] (Table 1).

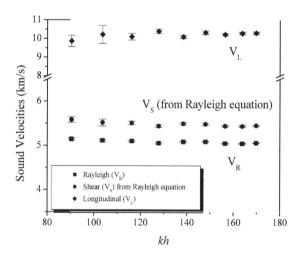

Figure 2. Velocities of the longitudinal (LM), shear and Rayleigh waves of nc-TiN/a-BN film (HF110602) versus kh ($kh = 2\pi h \Delta v/V$; h is the thickness of the film) measured by SBS.

Knowing the longitudinal wave velocity (V_L) and the Rayleigh surface wave velocity (V_R), it is possible to derive transverse velocity (V_S) from a Raylelgh dispersion equation for SAW in isotropic half-space [4]. The values of Rayleigh wave, shear wave (calculated) and longitudinal wave velocities along with Vickers hardness (H), shear (μ), bulk (K_S), Young's (E) adiabatic moduli, and the Poisson's ratio (σ) are presented in Table 1. All calculations related to the determination of elastic moduli assume the density of 5.4 g/cm^3.

Because the wavelength of the SAW is much lower than the thickness of the investigated films the SBS measurements of steel substrate are not required for the determination of elastic moduli. As seen from Fig. 2, measured sound velocities show no significant angular dispersion and therefore the film can be considered to be homogeneous with depth. Velocities of the acoustic waves evaluated by taking the average of the corresponding data points are shown in Table 1. In four nanocomposite films (see Table 1), a second peak (Fig. 1) was detected on the SBS spectrum. This peak may be attributed to the longitudinal lateral leaky wave [6].

Table 1. Characterization of nc-TiN/a-BN nanocomposite films by SBL spectroscopy.

Sample	h (μm)	V_R (km/s)	V_S (km/s)	V_L (km/s)	μ (GPa)	K (GPa)	E (GPa)	σ	H (GPa)
nc-TiN/a-BN (HF120602)	9.5	5.24(8)	5.82(9)		183(6)				
nc-TiN/a-BN (HF130602)	10.6	5.01(4)	5.45(4)	9.55(9)	160(2)	280(8)	403(7)	0.26(2)	
nc-TiN/a-BN (HF110602)	7.4	5.08(3)	5.48(5)	10.2(1)	162(3)	351(13)	421(4)	0.30(1)	
nc-TiN/a-BN[16]	6.1	4.90	5.32	9.40	152.8	270	386	0.264	46.4(4.8)
nc-TiN/a-SiN[16]	7.8	4.64	5.02	8.98	136	250	347	0.272	39.8(1.7)
TiN[10]		5.22	5.72	9.4	173	238	418	0.208	
c-BC$_2$N[17]			8.41	13.09	238	259		0.149	76

Superhard BC$_2$N nano-phase

Among the cubic B-C-N phases [18], the cubic BC$_2$N phase (c-BC$_2$N) reported by Solozhenko et al. [19] has novel and intriguing mechanical properties: its hardness is higher than that of cBN[1]. The Vickers hardness of nanocrystalline c-BC$_2$N (76 GPa) has been measured to lie between the hardness of diamond (115 GPa) and that of c-BN (62 GPa). Based on the nanohardness measurements, the shear modulus of the c-BC$_2$N has been predicted to be 447 GPa [1], which is even higher than that of diamond.

We have measured the elastic properties of nanocrystalline c-BC$_2$N phase by Brillouin scattering technique[17]. To estimate the average grain size of the c-BC$_2$N phase we conducted atomic force microscope (AFM) measurements. AFM images of the surface of the c-BC$_2$N phase are presented in Fig. 3. The surface nanostructure was studied in the area of the shallow notch of the specimen by the AFM (Nanoscope III, Digital Instruments, USA). AFM in contact imaging mode in the air was used to obtain constant-force topographic images. Both a V-shaped silicon nitride cantilever and a silicon type were used in this study, where the former has a pyramidal tip of 20 nm in the radius curvature, and the latter a small radius ($\sim$ 10 nm) of the tip. The images were taken in contact mode using Si$_3$N$_4$ cantilever. Grains of approximately 200 nm (189 $\pm$ 66 nm) are clearly seen in the AFM images (Fig 3). Average grain size of the grains measured by

AFM is nearly ten times larger than the value of the grain size obtained from TEM measurements [19].

Figure 3. Contact mode AFM images of the nanostructured c-BC_2N sample: scan size 2 μm, data scale (height) 195 nm.

DISCUSSION

The SBS measurements on nc-TiN/a-BN, nc-TiN/a-Si$_3$N$_4$ films show that shear and longitudinal velocities of these nanocomposite films are close to those of TiN (Table 1). It is also obvious from Table 1 that Young's moduli of the nc-TiN/a-BN nanocomposite films (403 GPa and 421 GPa) are lower than those measured by indentation technique (607 GPa). Further, the Young's moduli determined by SBS are lower than those measured for the films by vibrating reed technique (420–460 GPa)[5]. This discrepancy may be attributed to lower accuracy of the vibrating-reed technique, where several parameters such as thicknesses of the film, and the substrate, must be known with high accuracy. For SBS measurements of the acoustic wave velocities, the experimental error is less than 1 %.

Fig. 4 is a plot of Vickers hardness vs. shear modulus for nc-TiN/a-BN and nc-TiN/a-Si$_3$N$_4$ films, c-BC$_2$N, and selected hard materials [20]. The experimental data in the review paper [20] can be best-fitted by a linear trend:

$$H_o = 0.1769\mu - 2.899, \qquad (2)$$

where H_o is the hardness of the crystalline material. Based on the measured shear μ moduli of the c-BC$_2$N (238 GPa) the hardness H_o from eq. (1) should be 39.2 GPa in contrast to the H_o value of 76 GPa reported by Solozhenko et $al.$ [1]. Thus the SBS measurements demonstrate that superhard nanomaterials deviate sufficiently from the linear trend (Fig. 4). We attribute the higher hardness values of the nanocomposite films to the nanostructure. For metals, the hardness increases with decreasing grain size as $1/\sqrt{a}$ (where a is the grain size). This phenomenon is called the Hall-Petch effect[21]. A relationship for the strengthening of materials by grain refinement is represented by the Hall-Petch formula:

$$H = H_o + \frac{K_H}{\sqrt{a}}, \qquad (3)$$

where H is the hardness of the granular material, and K_H is a constant. Simple extension of the Hall-Petch dependence on the nanoscale grain size gives extremely high hardness. It has been reported recently that the hardness of nanocrystalline materials initially increased according to the Hall-Petch relation with decreasing crystallite size but decreased below 10-20 nm ("reverse Hall-Petch" effect) due to grain boundary sliding [12].

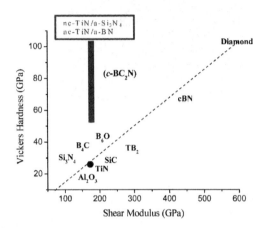

Figure 4. Vickers hardness vs. shear modulus for various hard and superhard crystalline materials: c-BC$_2$N, nc-TiN/a-Si$_3$N$_4$, nc-TiN/-a-Si$_3$N$_4$, and selected hard and superhard phases. Data for c-BC$_2$N are taken from Refs. [17,19]; for nc-TiN/a-Si$_3$N$_4$ from Refs.[22]; for the rest of the materials from Ref. [20].

The c-BC$_2$N phase has intriguing mechanical properties: its hardness is higher than that of cBN, but its elastic moduli are lower than those of cBN (see Table 1). The shear modulus value of 447 GPa for the c-BC$_2$N evaluated from the hardness measurement [1] is most likely overestimated. The nature of the high hardness of c-BC$_2$N is not understood yet. The hardness of superhard materials is mainly determined by atomic structure (bond density, bond length, and degree of covalent bonding) of the materials [20], leading to a correlation between hardness and value of shear modulus [20]. It is evident that the new nano-phase c-BC$_2$N deviates from the relation characteristic for superhard ($H>40$ GPa) materials (see Fig. 4). The other factor that can significantly increase hardness may be related to the granular structure. The combination of the SBS and AFM measurements enables us to determine parameters of the Hall-Petch equation. Using grain size ($a = 200$ nm) from AFM measurements, value of the hardness H_o (39.3 GPa) from eq. (1), and measured hardness H (76 GPa), we conclude that the coefficient of proportionality, K_H, in the eq. (2) should be 520 GPa nm$^{1/2}$. To our knowledge K_H has not been measured for superhard materials such as diamond and cBN; however, it has been determined for several hard materials such as TiN nanocoating (128 GPa $nm^{1/2}$)[23] and TiAl-based alloys (48 GPa $nm^{1/2}$) [24]. We note that the constant K_H of the BC_2N phase is nearly 10 times that of TiN nanocoating and 8 times that of TiAl-based alloys. Recently, Gao et al. [25] explained the high value of the c-BC_2N hardness using a semi-empirical model, which takes into account the bond density, bond length, and degree of covalent bonding of the covalent crystal c-BC_2N.

CONCLUSIONS

The SBS measurements demonstrate that superhard nanomaterials deviate from the H_0 vs μ linear trend. The hardness values of the nanocomposite nc-TiN/a-Si_3N_4 and nc-TiN/a-BN films are determined by the nanostructure but the relationship analogous to Hall-Petch effect is yet to be established. For the new superhard BC_2N phase, the hardness is likely determined by the atomic structure.

The SBS measurements also show that determination of the elastic moduli from hardness measurements gives higher values of the moduli than those measured by SBS. It supports the explanation made in Ref.[9], attributing the high value of the Young's modulus of the superhard nanocomposites to stiffening of the material due to high pressure in the vicinity of the tip.

ACKNOWLEDGMENTS

This work was supported by U.S. Army Contract No. DAAD19-00-1-0569 and partially by the National Science Foundation, Grant No. DMR-0102215. SOEST contribution number is 6703.

REFERENCES

[1] V. L. Solozhenko, S. N. Dub, and N. V. Novikov, "Mechanical properties of cubic BC2N, a new superhard phase", Diam. Relat. Mater., 10, 2228 (2001).

[2] M. Popov, M. Kyotani, R. J. Nemanich, and Y. Koga, "Superhard phase composed of single-wall carbon nanotubes", Phys. Rev. B, 65 (2002).

[3] S. Veprek and M. Jilek, "Super- and ultrahard nanocomposite coatings: generic concept for their preparation, properties and industrial applications", Vacuum, 67, 443 (2002).

[4] P. V. Zinin, in Handbook of Elastic Properties of Solids, Liquids, and Gases. Volume I: Dynamic Methods for Measuring the Elastic Properties of Solids, edited by M. Levy, H. Bass, R. Stern et al. (Academic Press, New York, 2001), Vol. 1, pp. 187.

[5] Z. S. Li, Q. F. Fang, S. Veprek, and S. Z. Li, "Evaluation of the internal friction and elastic modulus of the superhard films", Mater. Sci. Eng. A, 370, 186 (2004).

[6] M. G. Beghi, A. G. Every, and P. V. Zinin, in Ultrasonic Nondestructive Evaluation: Engineering and Biological Material Characterization, edited by T. Kundu (CRC Press, Boca Raton, 2004), pp. 581.

[7] M. H. Manghnani, S. Tkachev, P. V. Zinin, X. Zhang, V. V. Brazhkin, A. G. Lyapin, and I. A. Trojan, "Elastic properties of superhard amorphous carbon pressure- synthesized from C_{60} by surface Brillouin scattering", Phys. Rev. B, 64, 121403 (2001).

[8] S. Veprek, "The search for novel, superhard materials", Journal of Vacuum Science & Technology A, 17, 2401 (1999); S. Veprek, S. Reiprich, and S. H. Li, "Superhard Nanocrystalline Composite-Materials - the TiN/Si3N4 System", Appl. Phy. Lett., 66, 2640 (1995); S. Veprek and S. Reiprich, "A concept for the design of novel superhard coatings", Thin Solid Films, 268, 64 (1995).

[9] S. Veprek and A. S. Argon, "Mechanical properties of superhard nanocomposites", *Surf. Coat. Technol.*, **140**, 173 (2001).

[10] O. Lefeuvre, P. Zinin, and G. A. D. Briggs, "Leaky surface waves propagating on a fast on slow system and the implications for material characterization", *Ultrasonics*, **36**, 229 (1998).

[11] M. H. Manghnani, in *The 5th NIRIM International Symposium on Advances Materials (ISAM' 98)* (National Institute For Research in Inorganic Materials, Chichester, 1998), pp. 73.

[12] J. Schiotz, F. D. Di Tolla, and K. Jacobsen, W., "Softening of nanocrystalline metals at very small grain sizes", *Nature*, **391**, 561 (1998).

[13] M. C. Gupta and A. L. Ruoff, "Static compression of silicon in the [100] and in the [111] directions", *J. Appl. Phys.*, **51**, 1072 (1980).

[14] J. P. Poirier, *Introduction to the Physics of the Earth's Interior*. (Cambridge University Press, Cambridge, 1991).

[15] W. C. Oliver and G. M. Pharr, "Measurement of hardness and elastic modulus by instrumented indentation: Advances in understanding and refinements to methodology", *J. Mater. Res.*, **19**, 3 (2004).

[16] M. H. Manghnani, S. N. Tkachev, P. V. Zinin, C. Glorieoux, P. Karvankova, and S. Veprek, "Elastic properties of nc-TiN/a-Si_3N_4 and nc-TiN/a-BN nanocomposite films by surface Brillouin scattering", *J. Appl. Phys.*, **97** (2005).

[17] S. N. Tkachev, V. L. Solozhenko, P. V. Zinin, M. H. Manghnani, and L. C. Ming, "The elastic moduli of the superhard cubic BC_2N phase by Brillouin scattering", *Phys. Rev. B*, **68**, 052104(3) (2003).

[18] E. Knittle, R. B. Kaner, R. Jeanloz, and M. L. Cohen, "High-pressure synthesis, characterization, and equation of state of cubic C-B-N solid solutions", *Phys. Rev. B*, **51**, 12149–12156 (1995); T. Komatsu, M. Samedima, T. Awano, Y. Kakadate, and S. Fujiwara, "Creation of superhard B-C-N heterodiamond using an advanced shock wave compression technology", *J. Mater. Processing Technol.*, **85**, 69 (1999).

[19] V. L. Solozhenko, D. Andrault, G. Fiquet, M. Mezouar, and D. C. Rubie, "Synthesis of superhard cubic BC_2N", *Appl. Phys. Lett.*, **78**, 1385 (2001).

[20] D. M. Teter, "Computational alchemy: The search for new superhard materials", *Mater. Res. Soc. Bull.*, **23**, 22 (1998).

[21] E. O. Hall, "The deformation and ageing of mild steel: III Discussion of results", *Proc. Phys. Soc. Lond. B*, **64**, 747 (1951); N. J. Petch, "The cleavage strength of polycrystals", *J. Iron Steel Inst.*, **174**, 25 (1953).

[22] S. Veprek, S. Mukherje, P. Karvankova, H.-D. Männling, J. L. He, J. Xu, J. Prochazka, A. S. Argon, A. S. Li, Q. F. Fang, S. Z. Li, M. H. Manghnani, T. S, and P. Zinin, in *Surface Engineering: Synthesis, Characterization and Applications*, edited by A. Kumar, W. Meng, Y.-T. Cheng et al. (MRS, New York, 2003), Vol. MRS 750, pp. 1.

[23] F. J. Espinoza-Beltran, O. Che-Soberanis, L. Garcia-Gonzalez, and J. Morales-Hernandez, "Effect of the substrate bias potential on crystalline grain size, intrinsic stress and hardness of vacuum arc evaporated TiN/c-Si coatings", *Thin Solid Films*, **437**, 170 (2003).

[24] R. Bohn, T. Klassen, and R. Bormann, "Room temperature mechanical behavior of silicon-doped TiAl alloys with grain sizes in the nano- and submicron-range", *Acta Materialia*, **49**, 299 (2001).

[25]   F. M. Gao, J. L. He, E. D. Wu, S. M. Liu, D. L. Yu, D. C. Li, S. Y. Zhang, and Y. J. Tian, "Hardness of covalent crystals", *Phys. Rev. Lett.*, **91**, 015502 (2003).

Nanocomposites
and Nanostructures

INITIAL INVESTIGATION OF NANO-TiC/Ni AND TiC/Ni₃Al CERMETS FOR SOFC INTERCONNECT APPLICATIONS[1]

Hua Xie and Rasit Koc
Southern Illinois University at Carbondale
Carbondale, IL 62901

ABSTRACT

The development of new interconnect materials with high electrical conductivity that can operate in both reducing and oxidizing atmospheres is very important. TiC based cermets combining the high electrical conductivity of metal and high stability of ceramic could be a superior interconnect material for SOFC. High purity, high surface area, nano-size, and low cost TiC powders were produced utilizing the carbon coated Ti containing precursors. TiC-Metal and TiC-Intermetallic composites were processed by wet milling and hot pressing. The oxidation resistance and electrical conductivity studies are presented to provide results for applications of theses materials as an interconnect for Low/Intermediate Temperature Solid Oxide Fuel Cells. The results obtained from SEM are presented and interpreted.

INTRODUCTION

Solid oxide fuel cells (SOFCs) are promising candidates for next generation energy conversion device due to their inherently high efficiencies and zero emissions. One of the challenges for advancing SOFC technology is to develop interconnect materials with adequate electrical conductivity and long-term stability at operating temperature.

The requirements for interconnect are most stringent among all SOFC components. It must have high electrical conductivity; it must be physically and chemically stable in both reducing and oxidizing atmosphere at operating temperature; thermal expansion coefficient (TEC) must match well with those of the electrodes and electrolyte; it must have excellent impermeability to prevent direct combination of fuel and oxidant; there must be no reaction or inter diffusion between interconnect and its adjacent components under operating conditions.

Currently, the most common interconnect material for SOFC operating at 1000°C is strontium doped lanthanum chromite, which has good electrical conductivity at high temperatures and good stability in both oxidizing and reducing atmospheres. However, its major drawbacks include reduction in electrical conductivity at low oxygen pressures[1] and difficulty of processing under conditions that are appropriate for other cell components. Recently the operating temperature dropped to below 800°C, which enables metallic interconnect to replace the Sr-doped LaCrO₃ one. The formation of volatile chromia on the surface of chromium-based metallic

---

[1] Partially supported by the Solid State Energy Conversion Alliance (SECA), U.S. Department of Energy, under grant No. DE-FC26-04NT42224

interconnect can lead to severe degradation of the electrochemical performance of SOFC by poisoning the cathode. The contact resistances between a metallic interconnect and the cathode become so dominant that the electrical efficiency of the stack drops greatly due to the high growth rate of a semi-conducting, or even insulating, oxide scale[2].

A TiC-Ni or TiC-Ni₃Al cermet is a promising candidate for SOFC interconnects. A summary of the properties of TiC, Ni, Ni₃Al and SOFC components is given in Table 1[3-14]. From the table, it is clear that TiC shows high thermal stability and electrical conductivity as high as metals at room temperature. Ni is selected because Ni is already there in the anode and the electrical conductivity of nickel oxides is very high compared with that of chromia at elevated temperatures[15]. Ni₃Al, when used as a binder phase to fabricate dense cermets, is reported to be able to improve the oxidation resistance of the composites[16]. The thermal expansion coefficient of the cermet can be easily tailored by adjusting the percentage content of the ceramic and binder phase so as to match with those of other components in the system[17]. The electrical conductivity of TiC-Ni-Ni₃Al cermets under oxidizing and reducing conditions at high temperature as well as the correlative oxidation behavior has never been reported in literature.

Table 1. Properties of TiC, Ni, Ni₃Al and SOFC Materials

Properties	TiC	Ni	Ni₃Al	Y-PSZ	LSM	LSC
Density, g/cm³	4.92	8.9	7.5			
Melting Point, °C	3065	1454	1390			
Thermal Expansion Coeff. 1/°C x 10⁻⁶	7.4	13.3	~15	10.3	12.4	11.1
Electrical Conductivity (RT) S/cm x 10⁴	3.3	11.8	2.6		low	very low

In this study on TiC-Ni-Ni₃Al cermets, the electrical conductivity and oxidation behavior at 800°C for a period of 100 hours were initially investigated in order to demonstrate the suitability of TiC-Ni-Ni₃Al cermet for SOFC interconnect application.

EXPERIMENTAL PROCEDURES

TiC powders were made from 29.5wt% carbon coated TiO_2 precursors synthesized using a patented process developed by Dr. Rasit Koc[3]. Then, TiC powders were mixed with 30 weight percent of Ni and Ni₃Al powders, respectively, using wet milling in WC (tungsten carbide) container. Some of the cermet samples were produced using pressureless sintering in our laboratory, and others were synthesized using hot pressing at Greenleaf Corp. Hot pressed samples were subjected to X-ray diffraction to ensure the chemical composition. Samples were ground or polished to eliminate the pores on surface to minimize surface area. Small bars (~2.5×2.5×12 mm³) were cut from the hot-pressed samples. Weight, dimension and density of the bars were measured. Four-wire electrical conductivity measurements were performed using a Linear Research Inc LR-700 AC Resistance Bridge operating at a frequency of 16 Hz. Electrode paste was not used because the cross section area of the bars are close to that of the tip of the resistance bridge and the electrode paste might affect the oxidation behavior of the sample. The bar samples were, one at a time, heated from room temperature (24°C) to 800°C at a speed of 5 to 7°C per minute using a Carbolite Model CTF 17/75/300 Tube Furnace. An S-type thermocouple was placed close to the sample to read temperature. The electrical resistance was recorded for 100 hours at 800°C. The data of electrical conductivity were then calculated and plotted using Microsoft Excel. Weight gains of the tested samples were measured. Oxidized samples were subjected to XRD to investigate the composition of the oxidation scale. Scanning electron micrographs of the polished cross-section of the oxidized specimens were taken to study the microstructure and measure the thickness of the oxidation scale. The specimens were mounted in epoxy resin for polishing and were coated with carbon for SEM.

RESULTS AND DISCUSSIONS

According to the XRD patterns shown in Figure 1, the hotpressed cermet samples contain only TiC and Ni or Ni₃Al, respectively, and no peaks of a third phase has been found. After the electrical conductivity test, peaks of TiO_2, NiO and $NiTiO_3$ were found in the XRD patterns of TiC-30wt%Ni, as shown in Figure 2. This demonstrates that, TiC and Ni at the surface of the tested specimen were oxidized, forming an oxidation scale consisting of TiO_2, NiO and $NiTiO_3$. Figure 3 shows the microstructure of cross section area of the tested specimen, from which the thickness of the oxidation layer can be read as about 65μm. For TiC-30wt%Ni₃Al, TiC and Ni at the surface of were also oxidized, but only peaks of TiO_2 and NiO has been identified. Although no peaks of Al_2O_3 has been found in the XRD patterns, Al_2O_3, according to the study of M. Haerig and S. Hofmann[18], could probably form during the oxidation of Ni₃Al. As can be seen in Figure 4(II), the oxidation scale consists of 2 layers and some dots showing the traces of diffusion. Further study of EDX is necessary to find out the exact composition of the oxidation layer. From Figure 4, the thickness of the oxide scale can be read as about 60μm. Both specimens didn't gain much weight after the tests as shown in Table 2, because the oxidation scale once formed will become a protective layer preventing the specimen from further oxidation.

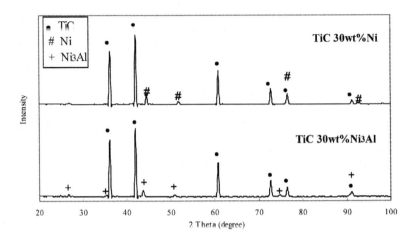

Figure 1. XRD of specimens (before electrical conductivity tests)

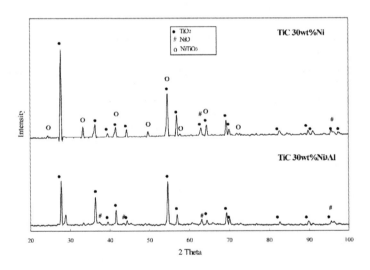

Figure 2. XRD of specimens (before electrical conductivity tests)

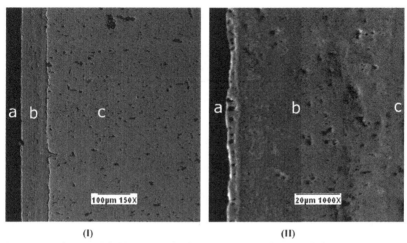

(I)   (II)

Figure 3. Backscattered electron images of polished cross section of tested TiC 30wt%Ni specimen, a) Epoxy Resin, b) Oxidation scale, c) Cermet

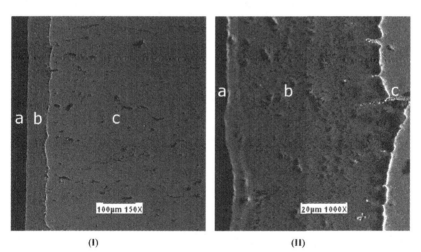

(I)   (II)

Figure 4. Backscattered electron images of polished cross section of tested TiC 30 wt% Ni$_3$Al specimen, a) Epoxy Resin, b) Oxidation scale, c) Cermet

Table 2. Weight gain of specimens after 100 hours electrical conductivity tests

Composition	Weight gain per surface area mg/cm^2	% Weight gain
TiC-30wt%Ni	6.56	3.1%
TiC-30wt%Ni₃Al	5.86	1.8%

Electrical conductivity σ in Siemens/cm (S/cm) was calculated using Equation 1.

$$\sigma = \frac{l}{RA} \quad (1).$$

Where l is the length of the specimen (in cm), A is the cross-section area of the specimen (in cm^2), and R is the resistance (in ohms) directly read from the resistance bridge. As can be seen from Figure 5 showing the electrical conductivity plotted as σ versus time in hours, the conductivity reached a maximum at 10-20 hours after the specimen was heated to 800°C and then started to decrease slowly. The decrease is believed to be due to the growth of the oxidation scale with relatively low conductivity, since the resistance read was a combination of that of the cermet and the oxides. Nevertheless, the conductivity remained at a very high level at the end of the 100 hours tests, compared to that of the cathode and other types of interconnect as shown in Table 3. Since, as previously mentioned, no electrode paste was used in order not to disturb the oxidation behavior of the specimens, the real contact area between the specimen and the tip of the resistance bridge, A' as shown in Figure 6, is smaller than the cross-section area A used in the conductivity calculation. Thus, the real electrical conductivity of the specimen should be even larger than the calculated one. Electrical conductivity measurements with the help of electrode paste on surface oxidized specimens and energy dispersive x-ray analyses will be taken in the future to further investigate the relation between the electrical conductivity and the composition of the oxide scale.

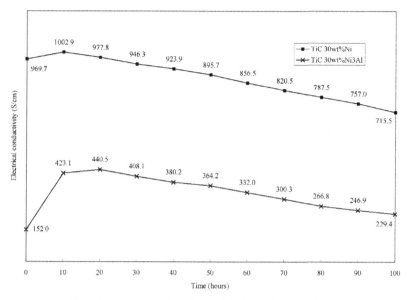

Figure 5. Electrical conductivity as a function of time at 800˚C for 100 hours.

Table 3. Comparison of the electrical conductivity of the cathode and different types of interconnect.

Composition	Electrical conductivity in oxidizing atmosphere (S/cm)	Temperature (˚C)
Sr-LaMnO$_3$ (Cathode)	150[19]	1000
LaCr$_{0.95}$Mg$_{0.05}$O$_3$	3.2[19]	1000
Stainless steel (Chromia)	Low because oxide scale dominates (1x10^{-2})[15]	800
TiC-30wt%Ni	700-1000	800

Figure 6. Schematic cross-section of the contact between the specimen and the tip.

CONCLUSION

The investigation confirms that TiC-Ni-Ni₃Al materials are more suitable to SOFC interconnect applications than metallic alloys. The electrical conductivity values of 700-1000 S/cm in oxidizing condition was recorded for dense materials. The reduction in electrical conductivity was believed to be due to the formation of oxides of Ti or Ni on the surface. These oxides once formed act as a protective layer in oxidation conditions which stops the further oxidation of the samples. They are stable in oxidizing conditions at 800°C; no degradation. TEC match is expected to be compatible with other cell components or it can be easily tailored.

ACKNOWLEDGEMENTS

The authors would like to thank Dr. Jason Goldsmith at Greenleaf Corp. for his help in sample preparation.

REFERENCES

[1] R. Koc and H. U. Anderson, "Electrical Conductivity and Seebeck Coefficient of (La, Ca) (Cr, Co)O₃," *Journal of Material Science*, **27** [20], 5477-5482 (1992).

[2] K. Hilpert, W. J. Quadakkers, and L. Singheiser, "Interconnects"; pp. 1037–1054, Vol. 4[8], in *Handbook of Fuel Cells – Fundamentals, Technology and Applications*. Edited by W. Vielstich et al., John Wiley & Sons, Ltd., Chichester, 2003.

[3] R. Koc and G. Glatzmaier, US Patent No: 5,417,952.

[4] S. Tasyuk and V.S.Nesphor, Poroshkovaya Metallurgiya, **8**, 627-630 (1987).

[5] B.Mei, R.Yuan, and X. Duan, "Investigation of Ni₃Al-matrix Composites Strengthened by TiC," *J. Mater. Res.*, **8** [11], 2830 (1993).

[6] R. Koc, C. Meng, and G. A. Swift, "Sintering Properties of Submicron TiC Powders from Carbon Coated Titania Precursor," *Journal of Materials Science*, **35**, 3131-3141 (2000).

[7] R.M. German, Liquid Phase Sintering, Plenum Press, New York,1985.

[8] R. Koc, and J.S. Folmer, "Synthesis of Submicrometer Titanium Carbide Powders," *J. Am. Ceram. Soc.*, **80**, 952-956 (1997).

[9] T.N. Tiegs, P.A. Menchhofer, P.F. Becher, C.B.Thomas, and P.K Liaw, "Comparison of Sintering Behavior and Properties of Aluminide-bonded Ceramics," Ceram. Engr. Sci. Proc., **19** [3], 447-455 (1999).

[10] C.T. Liu and J.O. Stiegler, "Ordered Intermetallics"; pp.913-942, Vol. 2, in *ASM Handbook*. Edited by J.R. Davis, ASM Internat., Metals Park, OH, 1990

[11] C.T.Liu and V.K Sikka, "Nickel aluminides forStructural Use," J. Metals, **38** [5] 19-21 (1986).

[12] A.M. Hodge and D.C. Dunand, , "Synthesis of Nickel–Aluminide Foams by Pack-aluminization of Nickel Foams," *Intermetallics*, **9**, 581-589 (2001).

[13] G. R. Stoeckinger and J. P. Neumann, "Determination of the Order in the Intermetallic Phase Ni$_3$Al as a Function of Temperature," *Journal of Applied Crystallography*, **3**, 32-38 (1970).

[14] C. Dimitrov, X. Zhang and O. Dimitrov, 1996, "Kinetics of Long-Range Order Relaxation in Ni$_3$Al: The Effect of Stoichiometry," *Acto mafer.*, **44** [4], 1691-1699 (1996).

[15] W.Z.Zhu and S.C. Deevi, , "Development of Interconnect Materials for Solid Oxide Fuel Cells," *Materials Science and Engineering*, A **348**, 227-243 (2003).

[16] P. F. Becher and K. P. Plucknett, "Properties of Ni$_3$Al-bonded Titanium Carbide Ceramics," *Journal of the European Ceramic Society*, **18** [4], 395-400 (1998).

[17] L.M. Zhang, J. Liu, R.Z. Yuan, and T. Hirai, "Properties of TiC-Ni$_3$Al Composites and Structural Optimization of TiC-Ni$_3$Al Functionally Gradient Materials," *Materials Science and Engineering*, **203**, 272-277 (1995).

[18] M. Haerig and S. Hofmann, "Mechanisms of Ni$_3$Al Oxidation between 500°C and 700°C," *Applied Surface Science*, **125**, 99-114 (1998).

[19] R. Koc and H.U. Anderson, "Investigation of Strontium-doped La(Cr,Mn)O$_3$ for Solid Oxide Fuel Cells," *Journal of Materials Science*, **27**, 5837-5843 (1992).

INTRA-TYPE NANOCOMPOSITES FOR STRENGTHENED AND TOUGHENED CERAMIC MATERIALS

Seong-min Choi, Sawao Honda, Shinobu Hashimoto, and Hideo Awaji

Nagoya Institute of Technology
Gokiso-cho, Showa-ku, Nagoya 466 8555, Japan
Corresponding author : choism1999@hotmail.com

ABSTRACT

We proposed a new concept of strengthened and toughened mechanisms in ceramic matrix nanocomposites. It was clarified that intra-type nanocomposites has a useful structure to improve fracture strength and fracture toughness of ceramic materials. In this work, we used α-alumina with average grain size of 260 μm and MoO_3 with purity 99.9 % as starting materials. We mixed alumina powder and neutralized MoO_3 solution with ethanol after dissolving MoO_3 in 28 % ammonia solution. We fabricated alumina-molybdenum nanocomposites using a PECS method. The results showed that the maximum values of fracture strength and fracture toughness of the specimens sintered at 1450°C were 860 MPa and 5.22 $MPa \cdot m^{1/2}$, respectively. The sintered specimen carried out heat-treatment at 800°C, 900°C, and 1000°C for 0 to 10 min in Ar atmosphere. The fracture toughness increased to 5.67 $MPa \cdot m^{1/2}$ after annealing at 800°C for 10 min because dislocations diffused in the matrix.

INTRODUCTION

It is desirable for ceramics in high-temperature structural applications to have improved mechanical performance. Nanocomposites have been showing a possibility for increasing mechanical properties. Niihara and his group proposed several nano-structures, such as intra-type, inter-type, intra/inter-type, and nano/nano-type, and reported many papers about nanocomposites[1-3]. Although, many research groups studied to improve the properties of ceramic materials[4-7], the strengthening and toughening mechanisms of nanocomposites were obscure. Recently, we noticed that an intra-type nano-structure is improved mechanical properties and proposed a new concept of strengthening and toughening mechanisms in nanocomposites. In order to clarify both the strengthening and toughening mechanisms, our group reported several results in ceramic-metal and ceramic-ceramic nanocomposites[8-10]. In this paper, we fabricated alumina-molybdenum(Mo) nanocomposites using a PECS (Pulse Electric Current Sintering) method and discussed the relationship between dislocations movement and improvement of mechanical properties.

THEORY

Toughening mechanism

Figure 1 shows a schematic illustration of the toughening mechanism of nanocomposites. Dispersed dislocations within the matrix grains after annealing for alumina/molybdenum nanocomposites are described in this figure. Sub-grain boundaries are generated around the nano-size molybdenum particles and these dislocations become sessile dislocations at room temperature, shown in Fig. 1(A). In this situation, when a tip of a propagating large crack reaches this area, these sessile dislocations in the matrix grains will operate as nano-crack nuclei in the vicinity of the propagating crack tip, shown in Fig. 1(B). The high stress-state in the FPZ is then released by nano-crack nucleation, and the nano-cracks expand the FPZ size and consequently enhance the fracture toughness of the materials.

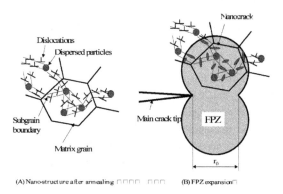

Fig. 1 Schematic description of the toughening mechanism in nanocomposites.

Strengthening mechanism

Sintered alumina has residual stresses in the grains and grain boundaries because of anisotropic thermal expansion, Young's modulus along the crystalline axes, and the crystallographic mis-orientation across the grain boundaries. In the sintered polycrystalline alumina, it is, therefore, conceivable that the large crack along a grain boundary created by synergetic effects of both residual stresses and processing defects will become equivalent to the grain size of the material and that the weakest crack generated along the boundary in the specimen dominates the strength of the specimen. Also, the fracture toughness of grain boundaries is usually lower than that of the matrix grains. Hence, polycrystalline alumina ceramics exhibit mainly intergranular fracture mode, as schematically shown in Fig. 2(A). Figure 3(A) shows the SEM (Scanning Electron Microscopy) observation of the fracture surface of monolithic alumina.

Nanocomposites, however, will yield dislocations around the particles, and the dislocation creation releases residual stresses in the matrix grains. Consequently, the defect size along the grain boundaries reduces in nanocomposites, as shown in Fig. 2(B). Also, the dislocations in ceramics are difficult to move in ceramics at room temperature, serve as origins of small stress concentrations, and create nano-cracks around the propagating crack tip. These nano-cracks slightly reduce the strength of the alumina matrix and the reduction of both the residual stress along the grain boundaries and the strength in the matrix is attributable to a change in the fracture mode from intergranular fracture in monolithic alumina to transgranular fracture in nanocomposites. Also, the fracture surface of the transgranular mode of nanocomposites is not a simple planar cleavage plane. Several steps are frequently observed on the surface and this phenomenon is likely to be evidence of nano-cracking in the wake of the FPZ. Figure 3(B) shows the SEM micrograph of the fracture surface of alumina/copper(Cu) nanocomposites fabricated by us. Reduction of both the defect size along the grain boundaries and the tensile residual stresses in the matrix grains due to dislocation activities results in improvement of the strength of nanocomposites.

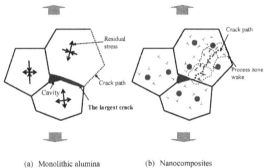

(a) Monolithic alumina   (b) Nanocomposites

Fig. 2 Schematic description of the strengthening mechanism in nanocomposites.

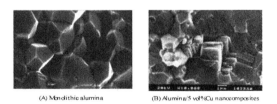

(A) Monolithic alumina   (B) Alumina/5 vol%Cu nanocomposites

Fig. 3 SEM micrographs of fracture surfaces of monolithic alumina (A) and alumina/copper nanocomposites (B).

EXPERIMENTAL PROCEDURE

Materials Preparation

We used α-Al_2O_3 (TM-D, Taimei Chem. Japan ; a mean grain size 260 nm) and MoO_3 (Mituwa Chem. Japan; purity 99.9%) as starting materials to fabricate Al_2O_3/Mo nanocomposites. We dissolved MoO_3 powder in 28 % ammonia solution and neutralized this solution with ethanol. We mixed α-Al_2O_3 and neutralized solution. In this time, we controlled Mo content from 1 to 5 vol%. The following Mo contents are apparent value. The mixtures are dried in evaporator at 70°C for 30 min and calcined at 550°C for 3 h. The calcined mixtures are reduced under high purity H_2 gas atmosphere at 900°C for 2 h. We sintered the reduced mixtures to a cylinder shape in a graphite die (inner size : 20 mm) at 1250°C, 1350°C, 1450°C, and 1550°C with heating rate of 50°C/min for 5 min under 30 MPa in vacuum atmosphere using a PECS method. The sintered specimens were cut into the sample dimension $2 \times 2 \times 10$ mm^3 and polished with 3 µm diamond paste.

Heat treatment and Characterization

The sintered specimen was heat treated at 800°X to 1000°C for 0 to 10 min in Ar atmosphere. The sintered and annealed specimens were measured the physical and mechanical properties. Bulk densities of the specimens after sintering were measured by the Archimedes' method. The fracture strength was obtained by a three-point bending test with a cross head speed of 0.5 mm/min and a span size of 8 mm for four specimens. The fracture toughness was measured using a single edged V-notch beam (SEVNB) method[11] with a notch depth of 1 mm for specimens with similar size of the fracture strength specimen. The SEVNB method uses a sharp V-shaped notch to estimate the intrinsic fracture toughness of the material. After a three-point bending test, the fracture surfaces were observed by SEM.

RESULTS AND DISCUSSION

Figure 4 shows XRD results of the α-Al_2O_3/Mo samples after (a) calcinations, (b) reduction, and (c) sintering. The calcined samples show α-Al_2O_3 and MoO_3 peak. But, reduced and sintered samples show α-Al_2O_3 and Mo metal peaks. From this result, it is known that molybdenum oxide was transformed to molybdenum metal due to the reduction in H_2 atmosphere and that we could

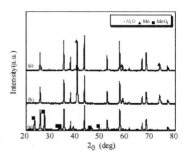

Fig. 4 XRD patterns of (a) calcined,
(b) reduced, and (c) sintered specimens

obtain α-Al₂O₃/Mo nanocomposites.

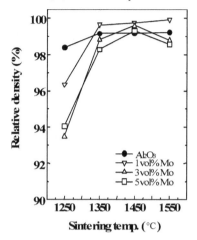

Fig. 5 Relative density of monolithic Al₂O₃ and α-Al₂O₃/Mo nanocomposites

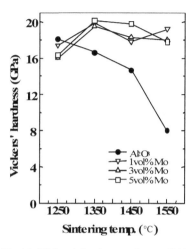

Fig. 6 Vickers' hardness of monolithic Al₂O₃ and α-Al₂O₃/Mo nanocomposites

Figure 5 shows the relative densities measured by Archimedes' method. In the case of monolithic alumina, we obtained dense alumina even at 1250°C. But, we could not obtain dense nanocomposites at 1250°C. From this result, it is known that nanocomposites should sintered above 1350°C to fabricate dense alumina/Mo nanocomposites.

Figure 6 shows the Vickers' hardness of monolithic alumina and nanocomposites. The Vickers' hardness of monolithic alumina at 1250°C is slightly higher than that of nanocomposites. The Vickers' hardness of monolithic alumina decreases with increasing sintering temperature because the grain size increased from 0.5 μm at 1250°C to 10 μm at 1550°C. But, in the case of nanocomposites, the Vickers' hardness shows higher than that of

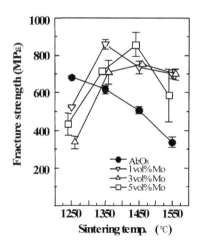

Fig. 7 Fracture strength of monolithic Al₂O₃ and α-Al₂O₃/Mo nanocomposites

monolithic alumina above 1350°C because Mo particles dispersed in alumina matrix decreased residual stresses of alumina in grains. The maximum value of Vickers' hardness is 20 GPa in nanocomposites.

Figure 7 shows the fracture strength of monolithic alumina and alumina/Mo nanocomposites measured by the three-point bending test. In the case of monolithic alumina, the fracture strength decreased with increased sintering temperature due to grain growth of alumina, and the value, 680 MPa at 1250°C is higher than that of nanocomposites. In contrast, the fracture strength of nanocomposites shows higher than that of monolithic alumina above 1350°C. The fracture strength of 1 vol% Mo nanocomposites increased to 860 MPa at 1350°C. But, this value decreased with increasing sintering temperature. In the case of 5 vol% Mo nanocomposites, maximum value is 860 MPa sintered at 1450°C and decreased at 1550°C. From these results, it is clear that the fracture strength of nanocomposites depends on the amount of Mo dispersed in the matrix.

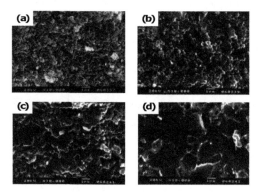

Fig. 8 SEM microphotographs of fracture surface of 5 vol% Mo nanocomposites
(a) 1250℃, (b) 1350℃, (c) 1450℃, and (d) 1550℃.

We observed microstructures of the fracture surface of 5 vol% Mo nanocomposites as shown in Fig. 8. The specimens sintered at 1250°C and 1350°C show inter-granular fracture mode. But, the fracture mode of nanocomposites above 1450°C was transformed from inter-granular mode to intra-granular mode. Although, the sample sintered at 1550°C shows intra-granular fracture mode, the fracture strength decreases comparing with the fracture strength of the specimen sintered at 1450°C because of grain growth.

Figure 9 shows the fracture toughness measured by SEVNB method. The tendency of fracture toughness with sintering temperature is similar to the fracture strength. The maximum value of the fracture toughness is 5.22 MPa·m$^{1/2}$ in 5 vol% Mo nanocomposites sintered at 1450°C. From this result, it is know that the optimum sintering temperature to generate dislocation in sintered specimen is 1450°C because the specimen shows the maximum values of the fracture strength and fracture toughness.

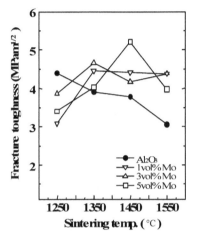

Fig. 9 Fracture toughness of monolithic Al$_2$O$_3$ and α-Al$_2$O$_3$/Mo nanocomposites

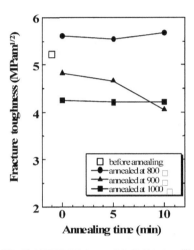

Fig. 10 Fracture toughness of α-Al$_2$O$_3$/ 5 vol% Mo nanocomposites sintered at 1450 °C after annealing

Figure 10 shows the fracture toughness of the specimen sintered at 1450°C after heat treatment at 800°C, 900°C, and 1000°C for 0 to 10 min in Ar atmosphere. We carried out heat treatment after sintering in order to diffuse dislocations in the matrix. From this result, the fracture toughness of the specimen annealed at 900°C, and 1000°C markedly decreased compared with that of the specimen before annealing. It is conceivable that the dislocations removed from the grains of alumina matrix. On the contrary, the fracture toughness of the specimen annealed at 800°C was higher than that of the specimen before annealing because the dislocations diffused in the matrix. The maximum value of the fracture toughness is 5.67 MPa·m$^{1/2}$ in the specimen annealed at 800°C for 10 min.

CONCLUSIONS

In this work, we used α-alumina and MoO_3 as starting materials. We mixed alumina powder and neutralized MoO_3 solution with ethanol after dissolving MoO_3 in 28 % ammonia solution. We fabricated Al_2O_3/Mo nanocomposites using a PECS method. The maximum values of the fracture strength and fracture toughness were 860 MPa and 5.22 MPa·m$^{1/2}$ sintered at 1450°C, respectively. The specimen sintered at 1450°C was heat-treated at 800°C, 900°C, and 1000°C for 0 to 10 min. The fracture toughness increased to 5.67 MPa·m$^{1/2}$ in the specimen annealed at 800°C for 10 min. From this result, it is clarified that appropriate heat treatment after sintering is important to improve the fracture toughness.

REFERENCE

[1]K. Niihara, "New design concept of structural ceramics" *J. Ceram. Soc. Japan*, 99, 974-82 (1991).

[2]M. Nawa, K. Yamazaki, T. Sekino, and K. Niihara, "Microstructure and mechanical behaviour of 3Y-TZP/Mo nanocomposites processing a novel interpenetrated intragranu ar microstructure" *J. Mater. Sci.* 31(11), 2849-58 (1996).

[3]Y. Hayashi, Y. H. Choa, J. P. Singh, and K. Niihara, "Mechanical and electrical properties of ZnO/Ag nanocomposites" *Ceramic Transactions*, 96, 209-18 (1999).

[4] M. Sternitzke, "Review : Structural Ceramic Nanocomposites" *J. Eur. Ceram. Soc.*, 17, 1061-1082 (1997).

[5]G. Pezzotti, T. Nishida and M. Sakai, "Physical limitations of the inherent toughness and strength in ceramic-ceramic and ceramic-metal nanocomposites" *J. Ceram. Soc. Jpn.*, 103 , 901–909 (1995).

[6]L.C. Stearns, J. Zhao and M.P. Harmer, "Processing and microstructure development in Al_2O_3-SiC 'nanocomposites'" *J. Eur. Ceram. Soc.* 10, 473–477 (1992).

[7]W. L. E. Wong, S. Karthik, and M. Gupta, "Development of hybrid Mg/Al2O3 composites with improved properties using microwave assisted rapid sintering route" *J. Mater. Sci.* 40(13), 3395-402 (2005).

[8]H. Awaji, S-M. Choi, and E. Yagi, "Mechanisms of toughening and strengthening in ceramic-based nanocomposites" *Mech. of Mater.*, 34, 411-22 (2002).

[9]T. Matsunaga, U. Leela-adisorn, Y. Kobayasi, S-M. Choi, and H. Awaji, "Fabrication of alumina-based toughened nanocomposites" *J. Ceram. Soc. Japan*, 113, 123-25 (2005).

[10]S-M. Choi and H. Awaji, "Nanocomposites-a new material design concept" *Sci. and Tech. of Adv. Mater.*, 6, 2-10 (2005).

[11]H. Awaji and Y. Sakaida, "V-notch technique for single-edge notched beam and chevron notch method" *J. Am. Ceram. Soc.*, 73, 3522-23 (1990).

PERIODIC NANOVOID STRUCTURE IN GLASS VIA FEMTOSECOND LASER IRRADIATION

Shingo Kanehira, Koji Fujita, Kazuyuki Hirao
Department of Material Chemistry, Graduate School of Engineering,
Kyoto University
Nishikyo-ku, Kyoto 615-8510, Japan

Jinhai Si
JST, Innovation Plaza, Kyoto
1-30 Ohara Goryo Nishikyo-ku, Kyoto 615-8245, Japan

Jianrong Qiu
Department of Materials, Zhejiang University
Hangzhou 310027, Zhejiang, China

ABSTRACT
We have observed periodically aligned nano-void structures inside a conventional borosilicate glass induced by a single femtosecond (fs) laser beam for the first time, to our knowledge. The spherical voids with nano-sized diameter were aligned spontaneously with a period along the propagation direction of the laser beam. The period, the number of voids, and the whole length of the aligned void structure were controlled by changing the laser power, the pulse number, and the position of the focal point.

INTRODUCTION

Ultrashort pulsed laser such as femtosecond (fs) laser are considered powerful tools for inducing nonlinear optical effects or microscopic modifications inside transparent materials due to their high power density of more than 10 TW/cm^2 and their ultrashort pulse width.[1-6] Using a focusing fs laser, it is possible to induce the multiphoton reduction of Au^{3+} or Ag$^+$ ions to their corresponding metal, the increase in refractive index, and the formation of void at the focal point inside the glass.[3]

A laser interference technique using fs laser pulses is applicable for the fabrication of periodic microstructures at a wide space of transparent materials since it produces a periodically modulated optical intensity with the size of the order of its wavelength.[7-10] Photonic crystals are materials in which the dielectric constant or refractive index is periodically modulated on a length scale comparable to the desired wavelength of the operation.[8,11] It is highly desired that periodic microstructures fabricated by a laser interference technique function as photonic crystals, because the fabrication time can be greatly shortened. Recently, we reported the fabrication of periodic microstructures in azodye-doped polymers by multibeam interference of fs laser pulses, in which a refractive index contrast (Δn) of about 5.5×10^{-4} was induced inside the polymer.[9,10] The value is large enough to induce the Bragg diffraction, but a larger Δn is needed for the periodic structures to function as photonic crystals. One method of achieving large Δn for photonic crystals is to make periodically aligned void structures in transparent materials. However, it is difficult to form a periodical void array using the laser interference technique in transparent materials such as glass.

In this paper, we report a novel method for fabricating periodic nano-sized voids inside a glass sample using a focused fs laser beam. In the conventional method, a focused fs laser beam produces a single void at the focal point.[4] Therefore, the fabrication of a periodic void array is achieved by precisely translating the glass sample.[12] Our method does not require such a procedure. In addition, periodically aligned voids are formed spontaneously with a period of micrometer length along the propagation direction of the laser beam. In addition, structural parameters of the void array, i.e., the period between voids, the number of voids, and the entire length of the aligned structure, can be controlled by adjusting the laser irradiation conditions. We will also discuss a possible formation mechanism of periodic void structures.

EXPERIMENTAL PROCEDURE

We used commercially available synthetic borosilicate glasses, S-1111 (Matsunami Co. Ltd., Osaka, Japan) of 12 mm × 6 mm × 0.9 mm in size. An optical absorption spectrum measured by a spectrophotometer (JASCO V-570, Tokyo, Japan) showed a small absorption over wavelengths of 350 nm to 1 μm. A regeneratively amplified 800 nm Ti:sapphire laser that emits 120 fs, 1kHz, mode-locked pulses was used in our experiments. The glass sample was put on a XYZ stage, which was controllable by a personal computer. A laser beam 5 mm in diameter in Gaussian mode was focused through a 100× objective lens with a numerical aperture (N.A.) of 0.9 in the interior of the borosilicate glass. The pulse energy of the fs laser at the sample location was approximately controlled by using neutral density (ND) filters between 10 and 40 μJ, which were inserted between the fs laser apparatus and microscope objective. An electronic shutter was used to control the number of pulses of the laser beams.

We observed the aligned void structure from a direction perpendicular to the incident laser beam through an optical microscope with a 150× objective lens using a color charge-coupled-device (CCD) camera. The morphology of the periodic voids and their surroundings was analyzed with high resolution using a scanning electron microscope (SEM, S-2600N, HITACHI Co, Ltd., Tokyo). The glass sample was cut to get the surface, including the periodic void structures, and the obtained surface was characterized by the secondary electron image.

RESULTS AND DISCUSSIONS

Figure 1 (a) shows a side view optical microscope photograph inside the borosilicate glass exposed to the fs laser beam at a pulse energy of 10 μJ. 250 pulses of the fs laser were launched in the interior of a glass sample 0.9 mm thick. The laser beam was focused at a depth of 750 μm from the entrance surface. The detailed experimental setup is illustrated in the left side of the optical microscope photograph. An aligned void structure is clearly observed below the focal point along the propagation direction of the fs laser beam. The formed voids have almost a spherical shape, and the neighboring two voids are independent of each other. No micro cracks or catastrophic collapses are observed around the voids or the focal point. An interesting feature is that the aligned void structure contains a region with periodically aligned voids (called the periodic part below) without connected voids or cracks, which are located at a distance of ~90μm

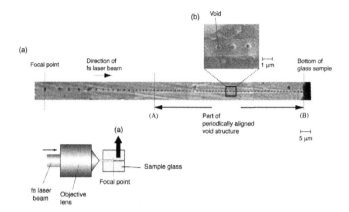

Figure 1. (a) Optical microscope photograph for a cross section of the 0.9-mm-thick borosilicate glass after irradiation with a 1-kHz fs laser beam with a pulse energy of 10 μJ for 1/4 s. Two hundred and fifty pulses were focused at a depth of 750 μm from the entrance surface. Laser pulses are incident from A to B. (b) An SEM photograph of periodic void structure.

from the bottom surface of the glass sample. As depicted in Figure 1 (a), a void with a diameter of 1.6 μm is formed at the focal point. The distance between the void at the focal point and the next void is 7.3 μm. Both the diameter of the voids and the interval between the two neighboring voids decrease gradually with the closing of the bottom surface of the glass sample, and approach limiting values at a distance of ~90 μm from the bottom surface. Namely, the periodic part exists at a range of ~90 μm from the bottom surface. The void size and intervoid separation in the periodic part are 380 nm and 1.7 μm, respectively. The diameter of the void confined at the focal point is slightly larger than the value of the lateral diffraction limit of the incident fs laser beam (~1.1 μm) estimated by the Rayleigh criterion: $\omega = 1.22\,\lambda\,/\,N.A.$, where λ is the laser wavelength of 800 nm and N.A. = 0.9. The discrepancy originates from the aberrations of the objective lens and has been often observed when the laser beam is deeply focused inside a glass sample.[13,14] Interestingly, the diameter of the voids in the periodical part is much smaller than that in the focal point, presumably due to a self-focusing effect.[15] We also confirmed that the use of an objective lens with a lower NA [0.45 (20×) or 0.3 (10×)] produced no void structures due to a lack of power density for dielectric breakdown at the focus. In addition, even when an objective lens with an N.A. of 0.9 was used, aligned void structure was not observed if fs laser pulses were focused near the entrance surface of the glass sample; instead, some cracks were formed below the focal point toward the bottom direction. Consequently, the deep focal point in the glass sample and the high N.A. of the objective lens are key points to the formation of periodic void structure.

Figure 1 (b) shows a SEM photograph of the periodic void structure near the bottom surface of the glass sample. A bright line 2 μm wide is clearly observed together with spherical voids. Formation of the bright line can be ascribed to the filamentation caused by irradiation with a focused fs laser beam. A filament phenomenon results from the dynamical competition between the self-focusing and defocusing of electron plasma. At powers around the critical value for self-focusing, the balance between self-focusing and defocusing of electron plasma can result in a long-range filament path.[16-19] Here, the filament path is always formed along the aligned voids, suggesting that the formation of the filament path is closely associated with the occurrence of aligned void structures. It should be noted that the formation of filament path in transparent materials using fs laser pulses has been examined from experimental and theoretical point of view, however, such a creation of periodical void structure has not demonstrated yet; this is the first observation, to our knowledge. A detailed explanation about the formation process of periodically aligned voids will be given below.

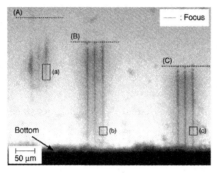

Figure 2. Optical microscope photograph of the cross section of the 1.1-mm-thick glass sample after irradiation with a 1-kHz fs laser beam with a pulse energy of 10 µJ for 1s. One-thousand pulses were focused at depths of (A) 750 µm, (B) 860 µm, and (C) 940 µm from entrance surface. Inset dotted lines show focal points of the fs laser beam. Enlarged parts of a - c in the upper photograph are depicted in the lower three photographs, respectively.

Figure 2 shows a side view optical microscope photograph inside the borosilicate glass irradiated with 1000 pulses of the fs laser. The glass sample had a thickness of 1.1 mm, which was thicker than the glass sample employed in Figure 1 (~ 0.9 mm in thickness). The pulse energy was 10 µJ, which was the same as in Figure 1. The laser beam was focused at depths of (A) 750 µm, (B) 860 µm, and (C) 940 µm from the entrance surface. In the case of (A), only an optical damaged line due to filaments is observed below the focal point, accompanying no aligned void structures, although the pulse energy and the position of the focal point relative to the entrance surface are the same as those shown in Figure 1. The difference in the experimental conditions between Figure 1 and Figure 2 (A) is the glass thickness: in the case of Figure 2 (A), the distance between the focal point and the bottom surface becomes long due to the increased glass thickness, and hence, the filament path does not reach the bottom surface. In both the cases of (B) and (C), on the other hand, the filament paths touch the bottom surface of the glass sample, and aligned void structures are successfully formed in the cores of the filament paths. The void size and period are almost identical for the two cases of (B) and (C). The entire length of the aligned void structure is 270 µm for (B) and 210 µm for (C), corresponding to the length of the filament path. Their aligned void structures are longer than those observed in Figure 1 (a) as a result of the increase of the filament path due to the increase in glass thickness. These results provide important evidence: aligned void structures can be formed only when the filament path reaches the bottom surface of the glass sample. They also show the possibility of adjusting the length of the aligned void structure by changing the length of the filament path.

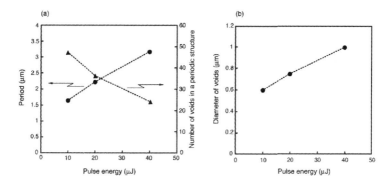

Figure 3. Number of voids and period (a), and diameter of voids (b) in area of periodically aligned void as a function of pulse energy of fs laser beam. 500 pulses of fs laser were launched in the interior of the glass sample 0.9 mm thick at depth of 750 μm from entrance surface. Pulse energy was varied from 10 to 40 μJ.

As mentioned above, the aligned void structure is composed of a periodic part near the bottom and a non-periodic part near the focal point. The length of the periodic part is independent of the pulse energy of the fs laser, while the number of voids and the interval between the two neighboring voids in the periodic part, i.e., the void period, vary depending on the pulse energy Both of the number of voids and the void period in the periodic part near the bottom are plotted in Figure 3 (a) against the pulse energy of the incident fs laser beam. The focal point was fixed 750 μm deep from the entrance surface of the glass sample 0.9 mm thick to form the aligned void structure. Pulse energy was varied between 10 and 40 μJ, and the number of pulses was fixed at 500. Preliminary experiments revealed that a long crack or catastrophic collapse was observed below the focal point at pulse energies higher than 40 μJ, while pulse energies less than 10 μJ led to only a visible damage line due to the refractive-index change. Considering these results, it is necessary to control the pulse energy between 10 and 40 μJ in borosilicate glass to induce periodic aligned voids. As shown in Figure 3 (a), the number of voids decreases with an increase in the pulse energy. For instance, the number of periodically aligned voids is 47 at a pulse energy of 10 μJ, while it is reduced to 25 when pulse energy is increased to 40 μJ. As naturally expected, the decreased number of voids leads to the increase in the void period because the length of the periodic part remains constant (~ 90μm) even if the pulse energy is varied; the void period gradually increases from 1.6 to 3.2 μm when the pulse energy is varied from 10 to 40 μJ. The void size in the periodic part also depends on the pulse energy [see Figure 3(b)]; upon irradiation with 1000 pulses, the void size increases from 600 nm to 1.0 μm as the pulse energy is increased from 10 to 40 μJ.

We presented a tentative explanation for the formation process of periodic void structures inside glass as shown in Figure 4. To create the periodic void structure shown in Figure 2, the fs laser beam must be focused at a deep position near the bottom surface of the glass sample; the filament path must also reach the bottom surface. In addition, the irradiation with one pulse of the focused fs laser does not induce periodic void structure, but produces a visible damage line due to the filament effect. These results indicate that sequential irradiation with some fs laser pulses, as well as the contact of the filament path with the bottom surface of the glass sample, is required for the formation of periodic void structure. Furthermore, we checked the morphology of the aligned void structure by changing the pulse number between 8 and 1000. When the pulse number is small, a few voids are formed around the bottom surface of the glass sample, indicating that void formation does not start from the focal point but from the bottom surface. As the pulse number is increased, void formation proceeds toward the focal point. The entire length of the periodic void structure gradually increases with increasing pulse numbers and exhibits a constant value of 130 ± 10 μm when the pulse number is increased above 125.

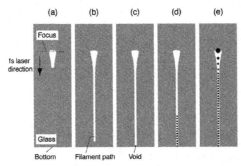

Figure 4. Schematic illustrations for the formation process of the periodic void structure. Void formation proceeds from Figure 4(a) - 4 (c).

Upon irradiation with a fs laser pulse, the refractive index change is induced at the focal point, as shown in Figure 4 (a), and then the filament path propagates toward the bottom surface of the glass sample, as shown in Figure 4 (b). Once the core of the filament line is raised to a sufficiently high temperature, it becomes highly absorbent. The fs laser light is more predominantly absorbed by the bottom surface to initiate microexplosion, since the threshold for dielectric breakdown that induces microexplosion is lower at the surface of the glass sample than in the inner regions. Namely, microexplosion takes place readily around the bottom surface to create a nano-sized void, as shown in Figure 4(c). When the next fs laser pulse propagates from the focal point, it is trapped in the heated region around the void formed beforehand, resulting in the production of a new high temperature region and the formation of the next void. The bright line around the voids, as shown in Figure 1 (b), i.e., the filament path, implies the localized melting of the glass sample due to increases in temperature caused by fs laser irradiation. The region that absorbs the laser light thus moves to a new point along the filament path, and the process is repeated many times, as shown in Figure 4 (d). Finally, the sequential void formation is completed at the focal point, as shown in Figure 4 (e). The fact that the void size at the focal point is different from that near the bottom surface is attributable to differences in the amount of light absorption that causes dielectric breakdown. As the pulse energy of the incident laser is increased, heating can occur even at the regions far from the void due to thermal diffusion, which causes an increase in the void period, as shown in Figure 3 (a).

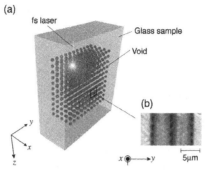

Figure 5. (a) Schematic illustration for 3D array of periodic void structures using our void-creating method, and (b) side view of periodic void structures using optical microscope with a 150× optical objective.

We have attempted to create a 3D periodic void structure inside the borosilicate glass, as schematically illustrated in Figure 5 (a). First, 1000 pulses of a fs laser with a pulse energy of 10 μJ were launched in the interior of a glass sample to form a periodically aligned void line along the z direction. Subsequently, the glass sample was translated to the x or y direction by 5 μm and irradiated with 1000 pulses of the fs laser. Repeating this process along the x and y directions can produce a 3D periodic void structure inside the glass sample. The side view of the 3D periodic void structure by optical microscope is shown in Figure 5 (b). Periodically aligned void structure with a void size of about 600 nm is clearly observed. Now we are trying to fabricate the nanostructure using this technique inside transparent materials with a high refractive index such as rutile TiO_2 crystals (refractive index $n = 2.4 - 2.6$, transparent in wavelength of 0.43 – 5.3 μm) and so on to achieve a large Δn and act as a photonic crystal.

CONCLUSIONS

We have found a novel method for producing periodic void structures in a borosilicate glass using fs laser pulses. The size of voids, and the period and the entire length of the void structure could be controlled by varying the pulse energy, pulse number, and focal point of the fs laser. Firstly, the filament path must reach the bottom surface of the glass sample to form the periodic void structure. Once one void is produced at the bottom surface of the glass sample, a sequent formation of many voids with a period occurs toward the focal point of the fs laser. The ability to easily fabricate such controllable periodic void structure guarantees applicability in optoelectronics areas, such as 3D photonic crystals.

ACKNOWLEDGMENT

We thank Prof. Peter G. Kazansky of Southampton University, UK, for his helpful suggestions and comments on this work. This work is partially supported by the Grant-in-Aid for Scientific Research (B) from the Ministry of Education, Culture, Sports, Science and Technology, Japan, and the Mitsubishi Foundation. The first author (S.K.) acknowledges the Research Fellowships of the Japan Society for the Promotion of Science.

REFERENCES

[1] K. M. Davis, K. Miura, N. Sugimoto, and K. Hirao, "Writing Waveguides in Glass with a Femtosecond Laser," *Opt. Lett.*, **21**, 1729-31 (1996).

[2] K. Miura, J. Qiu, H. Inouye, T. Mitsuyu, and K. Hirao, "Photowritten optical Waveguides in Various Glasses with Ultrashort Pulse Laser," *Appl. Phys. Lett.*, **71**, 3329-31 (1997).

[3] E. N. Glezer, M. Milosavljevic, L. Huang, R. J. Finlay, T. –H. Her, J. P. Callan, and E. Mazur, " Three-dimensional Optical Storage inside Transparent Materials," *Opt. Lett.*, **21**, 2023-25 (1996).

[4] E. N. Glezer and E. Mazur, "Ultrafast-laser Driven Micro-Explosions in Transparent Materials," *Appl. Phys. Lett.*, **71**, 882-84 (1997).

[5] J. Qiu, M. Shirai, T. Nakaya, J. Si, X. Jiang, C. Zhu, and K. Hirao, "Space-selective Precipitation of Metal Nanoparticles inside Glasses," *Appl. Phys. Lett.*, **81**, 3040-42 (2002).

[6] J. Qiu, X. Jiang, C. Zhu, M. Shirai, J. Si, N. Jiang, and K. Hirao, "Manipulation of Gold Nanoparticles inside Transparent Materials," *Angew. Chem. Int. Ed.,* **43**, 2230-34 (2004).

[7] K. Kawamura, N. Sarukura, M. Hirano, and H. Hosono, "Holographic Encoding of Permanent Gratings Embedded in Diamond by Two Beam Interference of a Single Femtosecond Near-Infrared Laser Pulse," *Jpn. J. Appl. Phys.*, **39**, L767-79 (2000).

[8] M. Champbell, D. N. Sharp, M. T. Harrison, R. G. Denning, and A. J. Turberfield, "Fabrication of Photonic Crystals for the Visible Spectrum by Holographic Lithography," *Nature*, **404**, 53-56 (2000).

[9] J. Si, J. Qiu, and K. Hirao, "Photofabrication of Periodic Microstructures in Azodye-Doped Polymers by Interference of Laser Beams," *Appl. Phys. B*, **75**, 847-51 (2002).

[10] J. Si, Z. Meng, S. Kanehira, J. Qiu, B. Hua, and K. Hirao, "Multiphoton-Induced Periodic Microstructures inside Bulk Azodye-Doped Polymers by Multibeam Laser Interference," *Chem. Phys. Lett.*, **399**, 276-79 (2004).

[11] E. Yablonovitch, "Photonic Crystals," *J. Mod. Opt.*, **41**, 173-94 (1994).

[12] H. –B. Sun, Y. Xu, S. Juodkazis, K. Sun, M. Watanabe, S. Matsuo, H. Misawa, and J. Nishii, "Arbitrary-Lattice Photonic Crystals Created by Multiphoton Microfabrication," *Opt. Lett.*, **26**, 325-27 (2001).

[13] C. B. Schaffer, A. Brodeur, J. F. Garcia, and E. Mazur, "Micromachining Bulk Glass by Use of Femtosecond Laser Pulses with Nanojoule Energy," *Opt. Lett.* **26**, 93-95 (2001).

[14]V. Koubassov, J. F. Laprise, F. Theberge, E. Forster, R. Sauerbrey, B. Muller, U. Glatzel, and S. L. Chin, "Ultrafast Laser-Induced Melting of Glass," *Appl. Phys. A*, **79**, 499-506 (2004).

[15]Y. R. Shen, *The principles of Nonlinear Optics*; John Wiley and Sons: New York **1984**, Ch. 17.

[16]A. Brodeur, F. A. Ilkov, and S. L. Chin, "Beam Filamentation and the White Light Continuum Divergence," *Opt. Commun.*, **129**, 193-98 (1996).

[17] Z. Wu, H. Jiang, Q. Sun, H. Yang, and Q. Gong, "Filamentation and Temporal Reshaping of a Femtosecond Pulse in Fused Silica," *Phys. Rev. A* **68**, 063820-28 (2003).

[18]L. Sudrie, A. Couairon, M. Franco, B. Lamouroux, B. Prade, S. Tzortzakis, and A. Mysyrowicz, "Femtosecond Laser-Induced Damage and Filamentary Propagation in Fused Silica," *Phys. Rev. Lett.,* **89**, 186601-04 (2002).

[19]Z. Wu, H. Jiang, L. Luo, H. Guo, H. Yang, and Q. Gong, "Multi Foci and a Long Filament Observed with Focused Femtosecond Pulse Propagation in Fused Silica," *Opt. Lett.* **27**, 448-50 (2002).

MATERIALS PROPERTIES OF NANO-SIZED FeAlN PARTICLES IN THIN FILMS

Yuandan Liu, R. E. Miller, Dingqiang Li, Qiquan Feng, W. Votava, Tao Zhang, L. N. Dunkleberger, X. W. Wang*
Alfred University
Alfred, NY 14802

R. Gray, T. Bibens J. Helfer
BTI
Rochester, NY 14586

K. Mooney, R. Nowak
SUNY Buffalo
Buffalo, NY 14260

P. Lubitz
Naval Research Laboratory
Washington, DC 20375

ABSTRACT

We report new results on materials properties of nano-sized FeAlN particles in thin films. Films were fabricated via a pulsed DC sputtering technique. Sputtering target materials were Fe_xAl_{1-x}, where x varies from 0.025 to 1. Film thickness varies from 0.1 to 3 micrometers, and particle size varies from 5 to 100 nanometers; depending on the fabrication conditions and target materials. Multi-layer films of FeAlN/AlN were also fabricated via the sputtering technique. Materials properties of the films were analyzed by SEM/EDS, TEM, and XPS. Magnetic properties of FeAlN thin films were also measured. Potential applications of such films in bio-medical fields will be discussed.

INTRODUCTION

Since 1990, nano-sized materials have generated substantial interests in many areas for applications.[1,2] In our previous work, nano-magnetic FeAl, FeAlO and FeAlN films with different Fe/Al ratios were fabricated via a sputtering technique.[3-7] It was observed that magnetic properties of the films are related to chemical composition, film thickness and fabrication conditions.

In contrast to bulk materials, a thin film material may exhibit different magnetic properties due to the constraint provided by the substrate.[7-9] The sputtering fabrication technique is utilized in this study to fabricate nano-magnetic FeAlN thin films. In particular, microstructural analysis of FeAlN thin films with different Fe/Al ratio are performed via TEM. The magnetic properties of FeAlN particles are also investigated. The intention of the current study is to search the applications of the materials in bio-medical fields. For example, a nano-magnetic material may exhibit larger coercive force than the corresponding bulk material.[10] In an MRI system, where the DC magnetic field is 1.5 Tesla or so, the enlarged coercive force is not a concern. The basic requirement for the MRI application is to have tunablility with magnetic properties. In this study, it will be interesting to see the relationship between the material properties of FeAlN particles in the films and their microstructures.

EXPERIMENTAL PROCEDURE

Nano-magnetic materials were fabricated by a PVD magnetron sputtering process. A Kurt J. Lesker Super System III deposition system outfitted with Lesker Torus magnetrons was utilized for the process[11] as shown in Figure 1. The vacuum chamber of the system is cylindrical, with a diameter of approximately one meter and a height of approximately 0.6 m. The base pressure is 2 µTorr. In these experiments, the targets are disks with a diameter of approximately 0.07-0.1 m. The sputtering gas is argon with a flow rate of 15-45 sccm. A pulsed DC power source is utilized at a power level of 500-2,000 W. The magnetron polarity switches from negative to positive at a frequency of 100 KHz, while the pulse width for the positive or negative duration can be adjusted to yield suitable sputtering results.

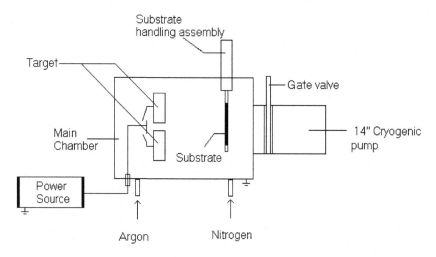

Figure 1. Schematic of the sputtering system. (Not to scale)

Weight ratios of the concentration of Fe and Al in the targets utilized are respectively: 11/89(W), 17/83(W) and 95/5(L).[12] Besides argon flowing at a rate of 15-45 sccm, nitrogen is supplied as a reactive gas with a flow rate of 15-50 sccm.

During fabrication, the pressure is maintained at 2-10 mTorr. This pressure range is found to be suitable for nano-magnetic material fabrication. The substrates are silicon wafers and metallic wires. The flat silicon wafers are bare, without a thermally grown silicon dioxide layer, and have a diameter of 0.1-0.15 m or less. The distance between the substrate and the target is 0.05-0.26 m. During the deposition, the wafer is fixed onto a substrate holder, without rotational motion. A typical wire is a copper wire with a diameter of approximately 0.5 mm. To deposit a film on a wire evenly, the wire is rotated at a rotational speed of 0.01-0.1 rps, and is moved slowly up and down along its symmetrical axis at a maximum speed of 0.01 m.s^{-1}. To achieve a film deposition rate on the flat wafer of 0.5 nm.s^{-1}, the power required for the AlN films and the FeAlN films is 500 W. A typical film thickness is between 100 nm and 1 μm with a typical deposition time between 200 and 2,000 s. In some film structure designs two layers are fabricated. The thickness and composition of the films is measured by a scanning electron microscope (FEI Quanta 200F ESEM) equipped with an energy dispersive X-ray system (Oxford Inca EDX Premium Si) or a scanning electron microscope (Philips 515 SEM) coupled with x-ray energy dispersive spectroscopy (Evex EDS analytical Si (Li)). To estimate the Fe/Al ratios, K-lines obtained in EDS are utilized. The phase formation studies are performed using transmission electron microscope (JEOL JEM-2000FX). The magnetic properties of films are determined by a superconductive quantum interference device (Quantum Design RF SQUID) magnetometer. A typical measurement procedure to obtain a hysteresis loop for a film is as follows: The sample is maintained at 5K, 20K, 50K, 100K and 300K, and initial magnetization is performed by applying a magnetic field from 0 to 2 Tesla. Thereafter, the field is changed from 2 Tesla to –2 Tesla and then backs to 2 Tesla to complete a hysteresis loop. The magnetization of the sample is plotted as function of the applied field.

RESULTS
SEM/EDS

The surface morphology of FeAlN thin film deposited on silicon wafers was observed by FEI Quanta 200F ESEM. The chemical compositions of thin films were analyzed by EDS with a spot size of approximately 1 μm. The size of an agglomerated FeAlN particle is approximately 100 nm. Figure 2 shows the EDS spectrum acquired from the top view of an FeAlN film (~200 nm thick, 11wt%Fe-89wt%Al target) deposited on a silicon wafer. (Due to the thin layer of 200 nm FeAlN, the signal from Si substrate is shown). Peaks of nitrogen, iron and aluminum are from the FeAlN film. The peak of carbon is presumably due to surface contamination and oxygen peak also is presumably due to surface contamination as well as surface oxidation during or after deposition. Peaks of gold and palladium are due to sample preparation. Figure 3 shows a cross-sectional view of an FeAlN film (11wt%Fe-89wt%Al target) with a film thickness of approximately 500nm.

In Figures 4 and 5, two SEM images of tilted fracture surface were obtained from a multi-layer film of (substrate)/FeAlN(1st layer/AlN(2nd layer)/FeAlN(3rd layer)/AlN(4th layer) (17wt%Fe-83wt%Al target) on a silicon wafer. These two images reveal that the FeAlN/AlN growth is columnized growth. The grain size of AlN layer (4th layer in Figure 4) is 30-200nm and the average diameter is about 100 nm, see Figure 5. Total thickness of the multi-layer film is approximately 3 μm. (The thickness of the 1st layer of FeAlN is approximately 1 μm.) EDS spectrum obtained from the cross-sectional view of the FeAlN (1st layer) of the four-layer film is shown in Figure 6. The peaks of Fe, Al and N are shown with the peaks of C, O, Au, Pd and Si, which are from contaminations, sample preparation and the substrate. SEM image of a cross-sectional view of this four-layer film is shown in Figure 7.

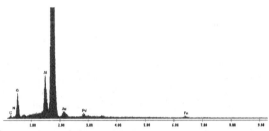

Figure 2. EDS spectrum of an FeAlN thin film (11wt%Fe-89wt%Al target) deposited on a silicon wafer.

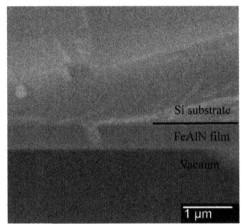

Figure 3. SEM image of a cross-section of an FeAlN thin film (11wt%Fe-89wt%Al target). From top to bottom, silicon wafer, FeAlN film and vacuum. The film thickness is approximately 500 nm.

Figure 4. SEM image of tilted fracture surface showing a four-layer film on a silicon wafer. From top to bottom, a surface of the film, the multi-layer film and the silicon wafer substrate are illustrated. Total thickness of the film is approximately 3 μm. The interface of 3rd layer and 4th layer is not clear.

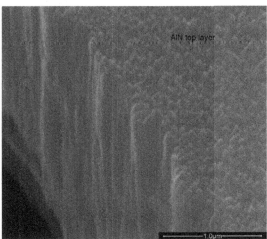

Figure 5. Enlarged SEM image of a four-layer film on a silicon wafer.

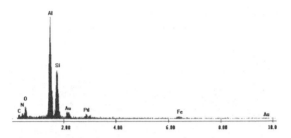

Figure 6. EDS spectrum obtained from FeAlN layer (1st layer).

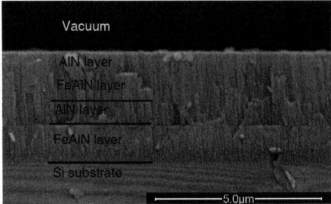

Figure 7. SEM cross-sectional image of a four-layer film on a silicon wafer substrate. From top to bottom, vacuum, AlN layer, FeAlN layer, AlN layer, FeAlN layer and silicon substrate. The interface of top AlN layer and FeAlN is not clear

The elemental concentrations in each layer of the film illustrated in Figure 7 are determined via K lines from EDS spectra, as shown in Table 1. Measured iron concentration is close to nominal iron concentration in the film according to calculated values 1 and 2 in Table I.

Table I. EDS results of FeAlN films/layer.

Layer	Target used (wt%Fe/wt%Al)	Measured thickness (nm)	Measured Fe concentration (wt%)	Measured Al concentration (wt%)	Measured Fe/Al ratio	Calculated value 1**	Calculated value 2***
1	17/83	1070	4.76	29.57	14/86	0.0462	0.0567
2	0/100	710	2.67	46.86	5/95	0.0119	0
3	17/83	890	6.31	41.88	13/87	0.0363	0.0471
4	0/100	540	4.95	45.40	10/90	0.0165	0
Total		3210				0.1110	0.1038

In Figure 8, an XPS spectrum of an FeAlN film deposited with 11wt%Fe-89wt%Al target is shown. Surface atomic concentrations of the films were analyzed with X-ray photoelectron spectrometry (XPS, PHI Quantera SXM). A typical area for analysis was approximately 3mm × 3mm. Aluminum, iron and/or nitrogen signatures are revealed. The surface concentrations are calculated using C (1s), N (1s), O (1s), Si (2p), Al (2p) and Fe (2p) photoelectrons, as shown in Table II.

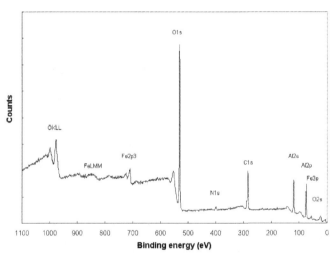

Figure 8. XPS spectrum of an FeAlN film (11wt%Fe-89wt%Al target) deposited on a silicon wafer.

Table II. XPS analysis of an FeAlN film (11wt%Fe-89wt%Al target) on a silicon wafer.

Sample	C(1s) (wt%)	N(1s) (wt%)	O(1s) (wt%)	Si(2p) (wt%)	Fe(2p) (wt%)	Al(2p) (wt%)
FeAlN	18.0	1.2	43.0	0	9.9	27.9

TEM/EDS

Samples for TEM analysis were made during film deposition. FeAlN films were deposited on a copper TEM grid. The nominal thickness of FeAlN was approximately 50 nm. The microstructures of the sample (without carbon coating) were studied using a Jeol JEM-2000FX TEM (Jeol, Japan) operated at 200 kV. Phase identification was performed by selected area diffraction (SAD), and chemical composition was obtained in the TEM, using a Princeton Gamma-Tech digital spectrometer and a PGT System4 Plus computer system (PGT, NJ).

Two different microstructures of FeAlN films were observed. Figure 9 show the TEM image of an FeAlN film deposited from 95wt%Fe-5wt%Al target. The structure of the crystallites in this film is circular and the diameter of the crystallites is approximately 2-5 nm. The chemical composition of an FeAlN film deposited from 95wt%Fe-5wt%Al target is shown in Figure 10. Only iron peak is associated with the film. There is an unidentified peak around 5.5keV in Figure 10. A different shape of the crystallites in an FeAlN film deposited from 11wt%Fe-89wt%Al target was observed, which shows needle-like structure. The width of the "needle" is less than 5 nm and the length of the "needle" is approximately 25 nm. The chemical composition of an FeAlN film deposited from 11wt%Fe-89wt%Al target is shown in Figure 11. The peaks of iron and aluminum are associated with FeAlN film. The copper peak is due to the copper grid. The peak between 0.35 keV and 0.55 keV may be due to nitrogen and/or oxygen. The exact atomic compositions are currently being studied.

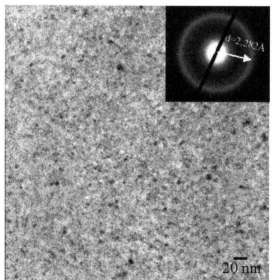

Figure 9. Bright Field (BF) TEM image of an FeAlN film (95wt%Fe-5wt%Al target) with SAD pattern insert.

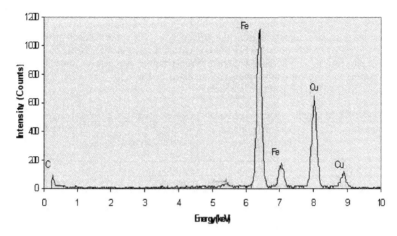

Figure 10. Chemical composition of an FeAlN film (95wt%Fe-5wt%Al target) on a copper grid.

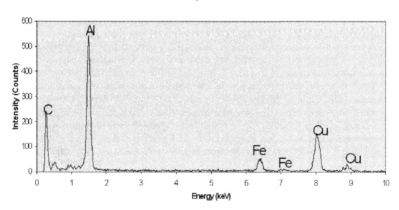

Figure 11. Chemical composition of an FeAlN film (11wt%Fe-89wt%Al target) on a copper grid.

SQUID

FeAlN powder for SQUID magnetization measurements was obtained from an FeAlN (11wt%Fe-89wt%Al target) coated silica bowl. FeAlN film was collected during film deposition and then the film was separated from the silicon bowl via a mechanical method. Magnetic hysteresis loops of FeAlN powder (11wt%Fe-89wt%Al target) at 5 K, 20 K, 50 K, 100 K and 300 K are plotted. The saturation and remnant magnetization, the coercive force H_C value of this FeAlN powder at different temperatures are tabulated in Table III. Saturation magnetization of this sample decreases as temperature increases.

Table III. Magnetic properties of FeAlN powder (11wt%Fe-89wt%Al target)

Temperature (K)	Magnetization at 2 T (emu/g)	Remnant magnetization (emu/g)	Coercive force (Oe)
5	1.66	6.09×10^{-4}	~5
20	6.28×10^{-1}	2.90×10^{-4}	~5
50	2.97×10^{-1}	4.65×10^{-4}	~12
100	1.66×10^{-1}	4.84×10^{-4}	~17
300	6.87×10^{-2}	5.07×10^{-4}	~18

DISCUSSION AND CONCLUSION

Nano-sized FeAlN particles were fabricated via a sputtering process. As revealed by SEM, the size of an agglomerated FeAlN particle is approximately 100 nm. The columnized growth structure of FeAlN/AlN multi-layer film was observed. The total thickness of a four-layer film is approximately 3 μm. XPS and EDS results reveal that the composition of the films is somewhat different from that of the target materials. Measured iron concentration is close to nominal iron concentration in the film via EDS analyses. XPS analysis shows that the iron concentration on the surface of the film is lower than the nominal value of the corresponding target. Two different microstructures in the films were observed according to TEM measurements. The shape of the crystallites in these two films is different. One is "needle-like" and the other is "circular-like". The size of a "needle-like" crystallite is approximately 3 nm× 25 nm. The diameter of a "circular-like" crystallite is approximately 2~5 nm. As measured by SQUID magnetization, saturation magnetization of FeAlN powders decreases as temperature increases. Currently, we are exploring the depth profile of the films.

ACKNOWLEDGEMENTS

Work partially supported by BTI, NYSTAR-CACT, and NSF-CGR.

FOOTNOTES

*. Corresponding author: Tel: +1-607-871-2130, Email address: fwangx@alfred.edu (X. W. Wang).

**. Measured wt%Fe/(measured wt%(Fe+Al)×(measured thickness/total thickness)).

***. Nominal wt%Fe/((nominal wt%(Fe+Al)×(measured thickness/total thickness)).

REFERENCES

[1]M. Solzi, M. Ghidini, and G. Asti, "Macroscopic Magnetic Properties of Nanostructured and Nanocomposite Systems," pp. 124-201 in *Magnetic Nanostructures.* Edited by S. Nalwa. American Scientific publishers, Stevenson Ranch, California, 2002.

[2]A. Carl and E. F. Wassermann, "Magnetic Nanostructures for Future Magnetic Data Storage: Fabrication and Quantitative Characterization by Magnetic Force Microscopy," pp. 59-92 in *Magnetic Nanostructures.* Edited by S. Nalwa. American Scientific publishers, Stevenson Ranch, California, 2002.

[3]Y. Liu, C. Daumont, E. Pavlina, R. E. Miller, C. McConville, X. Wang, R. W. Gray, J. L. Helfer, K. P. Mooney, and P. Lubitz, "Magnetic Properties of FeAlN Thin Films with Nano-sized Particles," *Proc. 9th Inter. Conf. on Ferrites (ICF 9),* 119-29 (2004).

[4]Y. Liu, X. Wang, R. E. Miller, L. N. Dunkleberger, R. W. Gray, J. L. Helfer, K. P. Mooney, and P. Lubitz, "Films with Nano-sized Particles Coated on Metallic Wires," *Proc. IWCS Symp.,* **53** 597-604 (2004).

[5]C. Daumont, Y. Liu, E. Pavlina, R. E. Miller, C. McConville, X. Wang, R. W. Gray, J. L. Helfer, K. P. Mooney, and P. Lubitz, "Characterization of FeAlN Thin Films with Nano-sized Particles," *Ceramic Transactions,* **159** 157-64 (2005).

[6]X. Wang, R. E. Miller, Y. Lin, R. W. Gray, J. L. Helfer, R. W. Nowak, and K. P. Mooney, "Nano-Magnetic Coatings on Metallic Wires," *Proc. IWCS Symp.,* **52** 647-53 (2003).

[7]X. W. Wang, R. E. Miller, P. Lubitz, F. J. Rachford, and J. H. Linn, "Nano-magnetic FeAl and FeAlN Films via Sputtering," *Ceramic Eng. & Sci. Proc.,* **24** [3] 629-36 (2003).

[8]G. G. Bush, "The Complex Permeability of a High Purity Yttrium Iron Garnet (YIG) Sputtered Thin Film," *J. Appl. Phys.,* **73** [10] 6310-1 (1993).

[9]X. W. Wang, R. E. Miller, J. H. Linn, and C. Washburn, "Technique Devised for Sputtering AlN Thin Films," *The GlassResearcher,* **12** [1-2] 35 (2003).

[10]M. DeMarco, X. W. Wang, R. L. Snyder, J. Simmins, S. Bayya, M. White, and M. J. Naughton, "Mossbauer and Magnetization Studies of Nickel Ferrites," *J. Appl. Phys.,* **73** [10] 6287-9 (1993).

[11]The Kurt J. Lesker Company, Clairton, PA. The magnetron is Torus 3 or 4.

[12]The suppliers of the targets are: W: Williams Puretek; L: Lesker Co.

PREPARATION AND PROPERTIES OF MULLITE - BASED IRON MULTI - FUNCTIONAL NANOCOMPOSITES

Hao WANG*, Weimin WANG, Zhengyi FU,
State Key Lab of Advanced Technology for Materials Synthesis and Processing,
122 Luoshi Road, Wuhan University of Technology,
Wuhan, Hubei Province, P. R. China, 430070

Tohru SEKINO and Koichi NIIHARA
Institute of Scientific and Industrial Research,
Mihogaoka 8-1, Osaka University,
Ibaraki, Osaka, Japan, 567-0047

ABSTRACT

Mullite-based nanocomposite powders with embedded iron nanoparticles were synthesized by reduction of sol-gel prepared $Al_{5.4}Fe_{0.6}Si_2O_{13}$ solid solution in hydrogen. Structural characterization studied by TEM revealed that the intra-granular iron nanoparticles along with inter type iron grains were obtained in mullite matrix. The magnetic properties of nanocomposite powders after reduction at 1200°C were measured by SQUID. It suggests that the intergranular and intragranular α-iron nanoparticles had the ferromagnetic and superparamagnetic behavior at room temperature, respectively. Dense Mullite-based iron composites were prepared from the nanocomposite powders reduced at different temperatures by hot pressing. The mechanical and magnetic properties were strongly affected by the reduction temperature. Flexural strength of mullite based iron nanocomposite reduced at 1200°C and successively hot pressed at 1650°C is one times higher than monothetic mullite. Furthermore, dense composites also behave ferromagnetic property at room temperature.

INTRODUCTION

Mullite ($3Al_2O_3 \cdot 2SiO_2$) is widely researched for high-temperature structural, electronic, and optical applications [1]. Compared with its outstanding high temperature strength and creep resistance, mullite exhibits low fracture toughness and strength at room temperature compared with other engineering ceramics [2]. To overcome this problem, some studies have been carried out by incorporating molybdenum particles in mullite matrix due to the similar thermal expansion coefficient values between them [3,4]. However, research on the influence of other metal inclusions on the properties of mullite composite is inadequate.

Based on the novel concept of the ceramic-based nanocomposites proposed by Niihara [5], techniques were developed to disperse Ni, Co, Mo, W, etc., nanoparticles into alumina or zirconia by reduction of metal compounds in ceramics matrix [6,7,8]. The mechanical properties of those nanocomposites are strongly enhanced, and other functions such as ferromagnetism can be provided by the metal nanoparticles, which shows the possible application of ceramic nanocomposites with multifunction. In contrast, nanocomposite powders with dispersion of metal nanoparticles can be obtained by the reduction of homogenous metal oxides solid solution [9,10], which also shows the possibility to gain metal particles dispersed composites after sintering. Mullite is a non-stoichiometric compound with the chemical formula of $Al^{VI}_2[Al_{2+2x}Si_{2-2x}]^{IV}O_{10-x}$ ($0 \leq x \leq 1$) [11]. Different amounts of transition metal cations mullite can be incorporated in mullite lattice by substitution of Al^{3+} or Si^{4+} at different positions. In the case of Fe-doped mullite, 12 wt% Fe_2O_3 can be dissolved in mullite matrix under 1200°C by replacing the Al^{3+} at octahedral site [12]. In the present study, mullite-based iron nanocomposite powders were prepared from the reduction of $Al_{5.4}Fe_{0.6}Si_2O_{13}$ solid solution at different temperatures. The dense composites were gained from the consolidation of nanocomposite powders by hot pressing at 1650°C. The formation, microstructure and magnetic properties of nanocomposite powders as well as the microstructure, mechanical and magnetic properties of the dense composites were investigated.

EXPERIMENTAL PROCEDURE
Sample preparation

To synthesize $Al_{5.4}Fe_{0.6}Si_2O_{13}$ solid solution, aluminium nitrate ($Al(NO_3)_3 \cdot 9H_2O$; High Purity Chemicals, 99.9%) and iron nitrate ($Fe(NO_3)_3 \cdot 9H_2O$; High Purity Chemicals, 99.9%) were used as the precursors for alumina and iron oxide respectively, while tetraethyl orthosilicate (TEOS, $Si(OC_2H_5)_4$; Wako, 95%) as the precursor for silica. The synthesis process was given in a previous study in detail [13]. Briefly, solid solution of $Al_{5.4}Fe_{0.6}Si_2O_{13}$ was prepared from the diphasic gel by sol-gel method and calcinations. To obtain the iron dispersed mullite nanocomposite powders, solid solution powders were reduced by flowing hydrogen gas from 1100°C to 1400°C with an interval of 100°C for 1 h.

The dense mullite-iron nanocomposite samples were prepared by reduction and hot pressing. Solid solution powders were dry ball milled for 24 h by alumina balls and screened through a No. 200 mesh sieve, then put into graphite dies. After hydrogen reduction from 1100°C to 1400°C with an interval of 100°C for 1 h with a heating rate of 20°C/min, the nanocomposite powders were successively consolidated at 1650°C under an applied load of 30 MPa in argon for 1 h by uniaxial hot pressing. The sintered bodies were cut and polished to the size of 3 ×4 ×40 mm for rectangular bars.

Characterization

The crystalline phases of composite powders before and after reduction were recorded between 10° to 90° at 4° 2θ/min by X-ray diffraction analysis (XRD; RU-200B, Rigaku Co. Ltd., Japan) using Cu Kα radiation. The powder XRD patterns for calculation of lattice parameters were measured using 0.5° 2θ/min with pure silicon as an internal standard. The microstructure of nanocomposite powders was observed by transmission electron microscope (TEM; H-8100T,

Hitachi Co. Ltd., Japan).

To investigate the microstructure configuration of dense composites, polished surfaces thermally etched at 1450°C for 1 h in argon atmosphere were observed using scanning electron microscope (FE-SEM; S-5000, Hitachi Co. Ltd., Japan). Vickers indentation, with a load of 49N for 15s, was used to determine the fracture toughness and hardness of composites. More than five independent measurements were performed for each experimental point. Young's modulus was tested using the resonance vibration technique. Flexural strength was measured by the three-point bending test with a span of 30 mm at a crossed speed of 0.5 mm/min at room temperature. At least five bars were tested for one experimental point. Magnetization measurements were conducted by means of a Quantum Design MPMS superconducting quantum interference device (SQUID) magnetometer.

RESULTS AND DISCUSSION
Nanocomposite Powders

Figure 1 shows the XRD profiles of samples before and after reduced at 1100-1400°C by hydrogen flow. Besides a small quantity of silica in some samples, the mullite and α-Fe are the only phases detected in all the samples. Because of the different stability of metal cations under reduction atmosphere in the solid solution, only Fe^{3+} can be reduced from the lattice of solid solution. The separation of iron phase from the lattice of mullite-iron oxide solid solution results in the decrease of lattice parameters and volume of unit cell. The lattice parameters and volume of unit cell for the samples reduced at different temperature are summarized in Table 1. The volume of unit cell sharply shrinks with the raise of reduction temperature till 1200°C and then keeps in stabilization. It reveals that the Fe^{3+} cation is completely transformed to iron phase when the reduction temperature exceeds 1200°C.

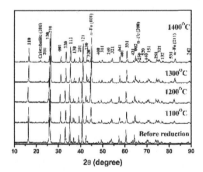

Figure 1. XRD profiles of samples before and after reduction in hydrogen flow at various temperatures.

Table 1. Summary of lattice parameters for the samples reduced at different temperatures in hydrogen flow.

Reduction temperature (°C)	a (Å)	b (Å)	c (Å)	Volume of unit cell (Å³)
Before reduction	7.5757±0.0035	7.7177±0.0044	2.9021±0.0019	169.678
1100	7.5437±0.0011	7.6926±0.0013	2.8847±0.0004	167.402
1200	7.5439±0.0013	7.6898±0.0012	2.8837±0.0004	167.287
1300	7.5413±0.0014	7.6909±0.0013	2.8842±0.0005	167.283
1400	7.5419±0.0012	7.6909±0.0013	2.8840±0.0005	167.284

The TEM images of reduced powders are shown in Figure 2. Two kinds of α-Fe nanoparticles with different particle size were observed. Some α-Fe grains with a size larger than 200nm are formed by the aggregation and growth of the iron nanoparticles located on/near the surface of the solid solution grain. α-Fe nanoparticles embedded in the mullite grains were also observed with a particle size less than 10nm. The formation of intra-granular nanoparticles is due to the precipitation of iron nanoparticles inside parent grains by changing the chemical state of iron through reduction. Mullite grain acts as a matrix to prevent the nanoparticles from further aggregation and growth.

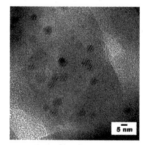

Figure 2. The TEM images of reduced powders

The magnetic hysteresis loops for mullite-iron nanocomposite powders reduced at 1200°C measured at temperature of 10 K (a) and 300 K (b) is shown in Figure 3. Saturated magnetizations (Ms), remanence ratio (Mr/Ms) and coercivity (Hc) of nanocomposite powders are 15.87 emu/g, 1.9%, 38Oe at 300 K and 16.86 emu/g, 3.0%, 45Oe at 10 K, respectively. It is known that for a superparamagnetic particle system, both the coercive field and remnant magnetization increase with decreasing temperature below the superparamagnetic-ferromagnetic transition temperature. And this increase is due to the lower thermal activation energy of spins at low temperature. In our results, the variation of coercivity and remanence ratio with temperature implies the existence of superparamagnetic iron nanoparticles in mullite-iron nanocomposite powders. As known, the magnetic behavior of metal nanoparticles is tightly related with their particle size. In the case of iron nanoparticles, the critical size of superparamagnetism is 14 nm [14]. From the observation of TEM, most of the intragranular iron nanoparticles are less than the critical size, which show the superparamagnetic behavior at room temperature. On the other hand, intergranular iron grains, with the size of hundreds of nanometers, are in the state of ferromagnetism.

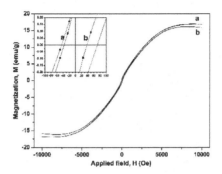

Figure 3. The hysteresis loop for $Al_{5.4}Fe_{0.6}Si_2O_{13}$ powders reduced at 1200°C. Insert shows enlargement of the plot near the origin ((a) at 10 K, (b) at 300 K).

Microstructural Characteristics and of Properties Sintered Bodies

The thermally etched surfaces of hot pressed samples and monolithic mullite are presented in Figure 4. Samples show the similar microstructure of some large acicular and rectangular grains dispersed in a matrix of fine equiaxed grains. Klug $et.$ $al.$ [14] reported that the single-phase mullite boundaries changed with temperature above ~1600°C and joined at ~1890°C at a composition 77 wt% Al_2O_3. Therefore, the single-phase mullite prepared from diphasic gel precursors will enter the mullite + liquid region of the phase diagram when heated between ~1650-1850°C. After the congregation and growth of embedded iron nanoparticles and iron grains through diffusion at high temperature sintering, the hybrid composite of mullite and iron was formed with iron particles locating in the triple join junctions (Figure 5).

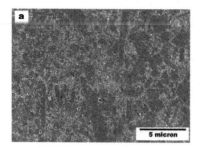 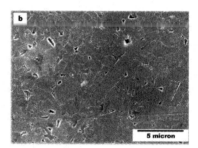

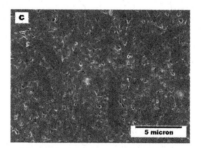

Fig. 4. SEM photograph for thermally etched surfaces of hot pressed samples.
(a: mullite; b: sample reduced at 1200 °C; c: sample reduced at 1300 °C)

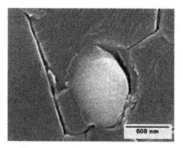

Fig. 5. Location of iron grain in dense composite

The mechanical properties of the studied specimens are given in Table 2. Flexural strength primarily rises to 400 MPa in the sample reduced at 1200°C and then sharply declines with the increase of reduction temperature, while the fracture toughness shows the maximum value in the sample reduced at 1300°C. Due to the incompletely transformation of Fe^{3+} to iron in the sample reduced at 1100°C, the σ_f is smaller than that of the sample reduced at 1200°C. When the reduction temperature exceeds 1200°C, all of the Fe^{5+} in solid solution is completely reduced. The iron phase is in the same content in composites. However, the grain size of iron phase is obviously related with reduction temperature. The iron grains grow large as the raise of reduction temperature. That is, the larger the iron grain size, the larger the critical flaw of composites. Thus the flexural strength declines as the raise of reduction temperature higher than 1200°C. The same result was reported in the research of mullite-Mo composites [3]. By the contribution of iron inclusion in composites, the toughness of composites slightly increases compared with monothetic mullite. The decline of Young's modulus and Vickers hardness are obviously resulted from the effect of soft metal phase.

Table 2. Mechanical and magnetic properties of Mullite-iron composites

	Flexural strength [MPa]	Indentation toughness [MPa·m$^{1/2}$]	Vickers hardness [GPa]	Young's modulus [MPa]
Mullite	207 ± 22	1.86 ± 0.06	12.30	226
Reduced at 1100°C	295 ± 11	2.05 ± 0.11	10.15	212
Reduced at 1200°C	405 ± 54	2.40 ± 0.13	10.82	213
Reduced at 1300°C	255 ± 18	2.03 ± 0.11	11.40	203
Reduced at 1400°C	204 ± 14	2.13 ± 0.22	11.34	211

The room temperature magnetic behavior of composite reduced at 1200°C is presented in Figure 6. Composite exhibits the typical ferromagnetism by the formation of hysteresis loop. Figure 7 shows the saturated magnetization and coercive force versus reduction temperature. The saturated magnetization of composites increased with reduction temperature till 1200°C. As the result of aggregation and growth of iron grain during sintering, the coercive force is around 20Oe.

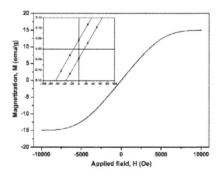

Figure 6. The hysteresis loop for $Al_{5.4}Fe_{0.6}Si_2O_{13}$ powders reduced at 1200°C and hot pressed at 1650°C. Insert shows enlargement of the plot near the origin.

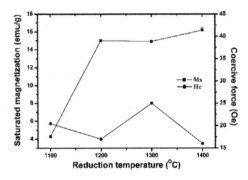

Figure 7. The saturated magnetizations and coercive forces of dense samples reduced at different temperature.

CONCLUSION

Dense mullite-based iron nanocomposites have been prepared through the reduction of $Al_{5.4}Fe_{0.6}Si_2O_{13}$ powders and successive hot pressing. Nanocomposite powders after reduction have an intra-type structure with the iron nanoparticles smaller than 10 nm embedded in mullite grains along with large iron grains ($\geq$ 200 nm). The mechanical and magnetic properties of dense composites are strongly affected by the reduction temperature. Flexural strength of mullite-based iron nanocomposite reduced at 1200°C and successively hot pressed at 1650°C is one times higher than monothetic mullite. Furthermore, dense composites also behave ferromagnetic property in room temperature.

ACKNOWLEDGEMENT
H. Wang thanks the National Natural Science of China (No. 50102003) in part for financial support.

REFERENCES
[1]1. A. Aksay, D. M. Dabbs and M. Sarikaya, "Mullite Processing, Structure, and Properies", *J. Am. Ceram. Soc.*, **74**, 2343- 58 (1991).

[2]T. –I. Mah and K. S. Mazdiyasni, "Mechanical Properties of Mullite", *J. Am. Ceram. Soc.*, **66**, 699-703 (1983).

[3]J. F. Bartolomeâ, M. Díaz, J. Requena, J. S. Moya and A. P. Tomsia, "Mullite/ Molybdenum Ceramic- Metal Composites", *Acta. Mater.*, **47**, 3891-99 (1999).

[4]J.´ F. Bartolomeâ, M. Díaz and J. S. Moya, "Influence of the Metal Particle Size on the Crack Growth Resistance in Mullite–Molybdenum Composites", *J. Am. Ceram. Soc.*, **85**, 2778- 84

(2002).

[5]K. Niihara, " New Design Concept for Structural Ceramics: Ceramic Nanocomposites", *J. Ceram. Soc. Jpn.*, **99**, 974-82 (1991).

[6]T. Sekino, T. Nakajima, S. Ueda, K. Niihara, "Reduction and Sintering of a Nickel-dispersed-alumina Composite and Its Properties", *J. Am. Ceram. Soc.*, **80**, 1139-48 (1997).

[7]T. Sekino, K. Niihara, "Fabrication and Mechanical Properties of Fine-tungsten-dispersed Alumina-based Composites", *J. Mater. Sci.*, **32**, 3943-49 (1997).

[8]H. Kondo, T. Sekino, N. Tanaka, T. Nakayama, T. Kusunose, K. Niihara, "Mechanical and Magnetic Properties of Novel Yttria-stabilized Tetragonal Zirconia/Ni Nanocomposite Prepared by the Modified Internal Reduction Method", *J. Am. Ceram. Soc.*, **88**, 1468- 73 (2005).

[9]V. Carles, Ch. Laurent, M. Brieu, A. Rousset, "Synthesis and characterization of Fe/Co/Ni alloys–MgO nanocomposite powders", *J. Mater. Chem.*, **9**, 1003-9 (1999).

[10]O. Quénard, E. D. Grave, Ch. Laurent, A. Rousset, " Synthesis, Charaterization and Thermal Behavior of $Fe_{0.65}Co_{0.35}$-$MgAl_2O_4$ and $Fe_{0.65}Ni_{0.35}$-$MgAl_2O_4$ Nanocomposite Powders", *J. Mater. Chem.*, 7, 2457- 67 (1997).

[11]W. E. Cameron, "Mullite: a Substituted Alumina", *Am. Min.*, **45**, 157-61 (1977).

[12]M. Ocaña, A. Caballero, T. González-Carreño, C.J. Serna, "Preparation by Pyrolysis of Aerosols and Structural Characterization of Fe-doped Mullite Powders", *Mater. Res. Bull.*, **35**, 775-88 (2000).

[13]H. Wang, T. Sekino and K. Niihara, "Magnetic Mullite-iron Composite Nanoparticles Prepared by Solid Solution Reduction", *Chem. Lett.*, **34**, 298-99 (2005).

[14] F. J. Klug, S. Prochazka and R. H. Doremus, "Alumina-silica phase diagram in the mullite region.", *J. Am. Ceram. Soc.*, **70**, 750- 59(1987).

[15]J. F. Löffler, J. P. Meier, B. Doudin, J. Ansermet and W. Wagner, "Random and exchange anisotropy in consolidated nanostructured Fe and Ni: Role of grain size and trace oxides on the magnetic properties", *Phys. Rev. B*, **57**, 2915-24 (1998).

* Author to whom correspondence should be addressed.

Dr. Hao WANG

State Key Lab of Advanced Materials Synthesis and Processing,

Wuhan University of Technology,

122 Luoshi Road, Wuhan 430070, Hubei Province, P. R. China

Tel: +86-27-8786-6762, Fax: +86-27-8721-5421

E-mail: shswangh@mail.whut.edu.cn

DESIGN OF NANOHYBRID MATERIALS WITH DUAL FUNCTIONS

Jin-Ho Choy
Center for Intelligent Nano-Bio Materials (CINBM), Division of Nanoscience and Department of Chemistry, Ewha Womans University, Seoul 120-750, Korea.

ABSTRACT

We have successfully synthesized inorganic-inorganic, organic-inorganic and bio-inorganic nanohybrids by applying an hybridization technique systematically to layered titanate, molybdenum disulfide (MoS_2), Bi-based cuprate superconductors ($Bi_2Sr_2Ca_{m-1}Cu_mO_y$ (m = 1, 2, and 3; BSCCO)), and to layered double hydroxides (LDHs). The inorganic-inorganic systems such as TiO_2-pillared titanate, TiO_2-pillared MoS_2, and CdS-MoS_2 hybrids were synthesized by exfoliation-reassembling method. A novel pillaring process of inorganic nanosol particles into layered compounds was developed to prepare inorganic-inorganic nanohybrids with a large surface area, high thermal stability, and enhanced photocatalytic activity. On the other hands, the organic-inorganic hybrids were achieved via intercalative complexation of iodine intercalated BSCOO with organic salt of pyridine. The high-T_c superconducting intercalate with its remarkable lattice expansion can be applied as a precursor for superconducting colloids when dispersed in an appropriate solvent. Especially, this organic-inorganic nanohybrid is expected to be a promising precursor for preparing the superconducting colloidal suspension, which could be applied to the fabrication of superconducting films or wires. Recently, we were very successful in demonstrating the formation of intercalative bio-inorganic hybrids. If necessary, LDH, as a reservoir, can be intentionally removed by dissolving it in an acidic media in such a way the interlayer biomolecules can be recovered or the intercalated biomolecules can be released from the LDH via ion-exchange reaction in electrolyte. It is, therefore, concluded that the inorganic LDH can play a role as a gene reservoir or carrier for various unstable organic or bio-molecules such as drugs and genes.

INTRODUCTION

Recently, inorganic-inorganic [1], organic-inorganic [2], and bio-inorganic [3] heterostructured nanohybrids with dual functions have attracted considerable research interests, due to their unusual physicochemical properties, which cannot be achieved by conventional solid state reactions. In order to develop new hybrid materials, various synthetic approaches, such as vacuum deposition, Langmuir-Blodgett technique, self-assembly, and intercalation method have been explored. Among them, the intercalation reaction technique—that is, the reversible insertion of guest species into two-dimensional host lattice—is expected to be one of the most effective ways of preparing new layered heterostructures because this process can provide a soft chemical way of hybridizing inorganic-inorganic, organic-inorganic, and biological-inorganic compounds. This field appears to be very creative, since it gives rise to an almost unlimited set of new compounds (hybrid compounds) with a large spectrum of known or unknown properties. As a consequence of the dual functionality of hybrid materials, this area is also a good field for scouting smart materials. For example, we were able to realize novel inorganic-inorganic hybrids with high photocatalytic activities via exfoliation-reassembling method. The most recent trend in the material chemistry goes toward the semiconductor-semiconductor hybrid in which both the components can be utilized in a photocatalytic process. The hybridization of two photoactive inorganic materials makes novel materials which can have bifunctional property and, furthermore, by coupling of two semiconductors, show much higher photoactivity than that of insulating host based system. In our coupled semiconductor nanohybrids, the improvement of efficiency is explained as the result of a vectorial transfer of electrons and holes from a semiconductor to another.

And we could also demonstrate new inorganic-organic hybrid systems with high-T_c superconducting properties. It is well known that the intercalation reaction occurs in highly anisotropic lamellar structures in which the interlayer binding forces are fairly weak, compared with the strong ionocovalent intralayer ones. The control of the strength of interlayer interactions makes it possible to probe the relation between interlayer coupling and superconductivity.

Finally, the layered double hydroxides (LDHs), so-called "anionic clays", have also received considerable attentions due to their technological importance in catalysis, separation technology, optics, medical sciences, and nanocomposite materials engineering. LDHs consist of positively charged metal hydroxide sheets, in which the interlayer anions (along with water) are stabilized in order to compensate the positive layer charges. The composition can be generally represented as $[M^{2+}_{1-x}M^{3+}_x(OH)_2][A^{n-}]_{x/n}mH_2O$, where M^{2+} is a divalent cation (Mg^{2+}, Ni^{2+}, Cu^{2+}, and Zn^{2+}), M^{3+} is a trivalent one (Al^{3+}, Cr^{3+}, Fe^{3+}, V^{3+}, and Ga^{3+}) and $A_{x/n}^{n-}$ is an exchangeable anion with charge n. The unique anion exchange capability of LDHs meets the first requirement as inorganic matrices for encapsulating functional biomolecules with negative charge in aqueous media. In this study, we present that biomolecules, such as DNA, ATP and antisenses etc. can be incorporated between hydroxide layers by a simple ion-exchange reaction to form bio-LDH nanohybrids. Moreover, the hydroxide layers can play a role not only as a reservoir to protect intercalated DNA, but also as a nonviral vector to transfer gene or drug to the cell.

EXPERIMENTAL SECTION
Inorganic-inorganic nanohybrids

As an example of inorganic-inorganic nanohybrid, TiO_2 (anatase)-pillared titanate was synthesized for the first time according to the following procedure [4]: the host caesium titanate, $Cs_{0.67}Ti_{1.83}\square_{0.17}O_4$, was prepared by heating a stoichiometric mixture of Cs_2CO_3 and TiO_2 at 800 °C for 20 h. The corresponding protonic form, $H_{0.67}Ti_{1.83}\square_{0.17}O_4{\cdot}H_2O$, was obtained by reacting the cesium titanate powder with 1 M HCl aqueous solution at room temperature for 3 days. During the proton exchange reaction, the HCl solution was replaced with a fresh one every day. The layered protonic titanate was exfoliated into single titanate sheets by intercalating the TBA (tetrabutylamine) molecule, as reported previously [5]. On the other hand, a monodispersed and non-aggregated TiO_2 nanosol was prepared by adding titanium isopropoxide (30 ml) with acetylacetone (20.38 ml) dropwisely to 0.015 M HNO_3 aqueous solution (180 ml) with vigorous stirring, and then by peptizing at 60 °C for 8 h. A TiO_2-pillared layered titanate nanohybrid was prepared by hybridizing the exfoliated layered titanate particles with the TiO_2 nanosol at 60 °C for 24 h. The resulting powder was collected by centrifuging (12 000 rpm, 10 min), washed with a mixed solution of distilled water and ethanol (1:1, v/v) to remove excess TiO_2 sol, and then dried in ambient atmosphere. Finally, the obtained material was heated at 300 °C for 2 h in order to complete the pillaring process.

As another example of inorganic-inorganic nanohybrid, TiO_2-pillared MoS_2 was synthesized by the exfoliation-reassembling method [6]. The host, molybdenum sulfide (2H-MoS_2), was lithiated by 3-fold molar excess of 1.6 M n-BuLi for three days to prepare $LiMoS_2$.

The product, $LiMoS_2$, was washed with n-hexane in a glove box and dried in vacuum. Subsequently, de-ionized and de-gassed water was added to $LiMoS_2$ to produce a suspension of exfoliated MoS_2 in a concentration of $1g/L$, and the suspension was sonicated for 10 min. The monodisperse non-aggregated nano-sol of titania were also prepared by the hydrolysis of titanium butoxide in the presence of acetylacetone and para-toluenesulfonic acid [7]. The exfoliated aqueous MoS_2 suspension was mixed with TiO_2 nano-sol in ethanol media (1:10 molar ratio of Ti to Mo). After stirring the mixture for 24 h, the flocculated product was centrifuged, washed several times with distilled water and ethanol (1:1, v/v) and dried in vacuum. $CdS-MoS_2$ hybrid was also prepared through the analogue procedure of TiO_2-pillared MoS_2 as reported previously [8].

Organic-inorganic nanohybrids

As a model system of organic-inorganic nanohybrid, the layered $Bi_2Sr_2Ca_{m-1}Cu_mO_y$ (m = 1 and 2) compounds as the inorganic host lattices were synthesized by conventional solid state reaction with nominal compositions of $Bi_2Sr_{1.6}La_{0.4}CuO_x$ (Bi2201) for m = 1 and $Bi_2Sr_{1.5}Ca_{1.5}Cu_2O_y$ (Bi2212) for m = 2, where the Sr ion is partially substituted by the La ion or Ca one to obtain single-phase samples. The intercalation of organic chain molecules into the pristine material was achieved by the following procedure: at first, the HgI_2-intercalated $Bi_2Sr_2Ca_{m-1}Cu_mO_y$ (m = 1 and 2; HgI_2-Bi2201 and HgI_2-Bi2212) compounds were prepared by heating the guest HgI_2 and the pristine materials in vacuum-sealed Pyrex tubes, as reported previously [2]. Then, the intercalation of organic chain molecules was carried out by the solvent-mediated reaction between HgI_2 intercalates and alkylpyridinium iodide. The reactants of Py-$C_nH_{2n+1}I$ (n = 1, 2, 4, 6, 8, 10, and 12) were obtained by reacting alkyliodide with 1 M equivalent of pyridine in diethylether solvent. The HgI_2 intercalates were mixed with two excess Py-$C_nH_{2n+1}I$, to which a small amount of dried acetone was added. Each solvent-containing mixture was reacted in a closed ampoule at 40 °C for 6 hours and washed with a solvent blend of acetone and diethylether (1:1 volumetric ratio) to remove the excess reactant of Py-$C_nH_{2n+1}I$. And finally the resulting products were dried in vacuum [2]. The superconducting colloidal suspension could be obtained only by sonicating the organic-salt intercalates in acetone solvent [9], which was then deposited on a substrate by electrophoretic deposition (EPD) technique and subsequent heating to obtain well aligned thin and thick films.

Bio-inorganic nanohybrids

Bio-inorganic nanohybrids have been prepared through intercalation route of biomolecules into inorganic lattice such as layered double hydroxides. The pristine Mg_2Al-NO_3-LDH was simply prepared by coprecipitation from aqueous solutions containing metal ions $(Mg(NO_3)_2 \cdot 6H_2O$ and $Al(NO_3)_3 \cdot 9H_2O$; 2:1 molar ratio) with dropwise titration of a base (NaOH) under nitrogen atmosphere. The resulting white precipitate was further aged for 24 hrs, then collected by centrifugation and washed thoroughly with decarbonated water. The biomolecule–LDH hybrids were then prepared by ion–exchanging the interlayer nitrate ions in the pristine LDH with various biomolecules such as herring testis DNA, or As-*myc* antisense oligonucleotide (As-*myc*; 5' d (AACGTTGAGGG GCAT) 3') at pH =7. The pristine LDH was dispersed in a deaerated aqueous solution containing an excess of dissolved, biomolecules and reacted for 48 h.

Electrophoretic analysis was carried out to prove that LDH lattice can play a role as a gene resovoir. In order to test DNase I resistance, 96 units of DNase I was added to the nanohybrids (8 μg) and the native DNA (15μg) directly and treated for a half, 1, and 24 hours at 37°C. For the recovery experiment of DNA from the hybrid, after an addition of DNase I to the hybrid for an hour and quenched with DNase I stop solution (0.2M NaCl, 40mM EDTA, and 1% SDS), then adjusted to pH 2 to dissolve the host lattice. Then DNase I treated samples and acid treated samples were analyzed through gel eletrophoresis.

The cellular uptake experiments were carried out for As-*myc*-LDH hybrid. HL-60 cells were also used to prove that the LDH could act as a drug delivery vector in gene therapy. HL-60 cells were exposed to As-myc or As-*myc*-LDH hybrid at a final concentration of 5, 10, 20 μM, respectively. Cell viability was estimated by spectrophotometry measurement of the samples treated with MTT assay. MTT assay is a colorimetric assay that measures the reduction of 3-(4,5-dimethylthiazol-2-yl)-2,5-diphenyl tetrazolium bromide (MTT reagent) by mitochondrial succinate dehydrogenase. Since the reduction of MTT can only occur in metabolically active cells, the level of activity is a measure of the viability of the cells.

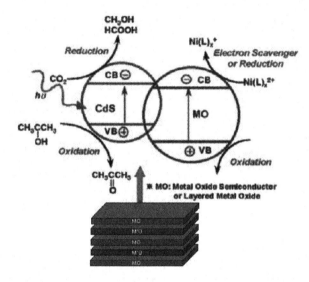

Figure 1. Schematic model for electronic and crystal structures for new hybrid photocatalysts

RESULTS AND DISCUSSION

Inorganic-inorganic nanohybrids

The nanosized MoS_2 with a narrow bandgap absorbs visible lights and can be coupled by CdS and TiO_2. The production of semiconductor nanoparticles and their organization on two dimensional solids are of great importance for the fabrication of nanostructured devices. The quantum effects of such particles would open a new way of potential application in photocatalytic and/or solar cell devices. The colloidal nanoparticles (for example, CdS, CdSe and TiO_2) and nanosheets (for example, exfoliated MoS_2 and layered titanate) can be prepared by intercalative exfoliation in an appropriate solvent. For example, the intercalation of TiO_2 and/or CdS nanocluster into the two dimensional MoS_2 lattice could be also demonstrated by exfoliating and reassembling the lithiated molybdenum disulfide ($LiMoS_2$) in the presence of cationic TiO_2 and/or CdS nano-sol particles in an aqueous solution, respectively, to obtain the semiconductor (UV)-semiconductor (Vis) nanohybrids (Figure 1).

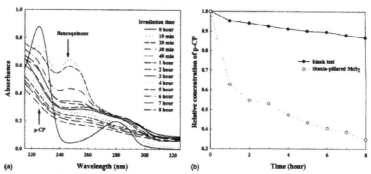

Figure 2. (a) UV spectra for degradation of 4-CP during the given time. (b) Relative concentration of 4-CP as a given time.

We explore the use of this high ordered hybrid as photocatalysts for degradation of 4-chlorophenol (4-CP) in water solution. UV light was used for the photocatalysis. When the 4-CP is photocatalytically decomposed, it finally becomes carbon dioxide via benzoquinone. These molecules, as well as the 4-CP itself, have characteristic optical absorption so that they can be identified by spectrometry. Quantitative data on the reaction products were obtained from measurements of transmitted intensity, assuming that Beers law applies. In particular, we used the absorption peaks at 225 nm for 4-CP and 245 nm for benzoquinone, respectively. This absorption drops under UV irradiation concomitantly with an increase of the absorption corresponding to benzoquinone. These observations are in general agreement with the mechanism of the decomposition of 4-CP. After having reached a maximum, the benzoquinone concentration decreases, which can be clearly seen from the data in Figure 2 (a) taken after 1 hour. The relative rate of the UV-induced degradation of 4-CP was determined by dividing the absorbance measured at 225 nm by the corresponding value obtained in the dark (i.e. at time zero). Figure 2 (b) shows relative concentration of 4-CP as a function of time. The concentration dramatically decreases after 1 hour of UV irradiation. It is 68% of the initial concentration after 1 hour. After that, the decomposition aspect displays a linear drop of the 4-CP concentration with time. And, finally, the concentration of 4-CP falls down up to 47% after 8 hour. From this photocatalytic experiment, we found that titania-pillared MoS_2 has a relatively high decomposition activity, especially effective at an early stage.

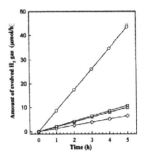

Figure 3. Cumulative amount of hydrogen gas evolved from 30 cm^3 solution of 0.1 M triethanolamine containing 10 mg of dispersed layered caesium titanate (squares), layered protonic titanate (diamonds), anatase TiO_2 nano-sol particles (triangles), and TiO_2-pillared layered titanate (circles), respectively.

On the other hands, the photocatalytic activity of TiO_2-pillared titanate was evaluated by measuring the total volume of hydrogen gas evolved during the irradiation of catalyst suspensions in water (Figure 3). Compared with the pristine layered caesium titanate (2.00 μM h^{-1}), the TiO_2-pillared titanate shows higher catalytic activity (8.68 μM h^{-1}), which is not only due to the presence of the anatase TiO_2 nano-sol particles stabilized in the interlayer space of layered titanate but also to the suppression of electron–hole recombination. It is worthwhile to note here that the photocatalytic activity result for the present samples was obtained in a Pyrex reactor with a very small capacity of 30 cm^3. Since the photocatalytic activity is dependent on the number of photoinduced electrons, which are proportional to the surface area of catalyst interacting with photons, a further study is therefore needed in order to enhance the photocatalytic activity by enlarging the reactor size (>1000 cm^3) and its design.

Organic-inorganic nanohybrids

We have successfully synthesized organic/inorganic nanohybrids by applying an intercalation technique systematically to Bi-based cuprate superconductors, $Bi_2Sr_2Ca_{m-1}Cu_mO_y$ (m = 1, 2, and 3; BSCCO).

The zero-field-cooled (ZFC) dc magnetizations of the pristine Bi2212 and intercalates were measured as a function of temperature (Figure 4A). In spite of a remarkable expansion of basal spacing, all of the organic intercalates exhibit superconductivity with an onset Tc of 81 to 82 K, which is higher than those for the iodine intercalate [10] ($Tc \approx$ 63 K, $\Delta d \approx$ 3.6) and the HgI_2 intercalate ($Tc \approx$ 68 K, $\Delta d \approx$ 7.2) and even slightly greater than that for the pristine material ($Tc \approx$ 78 K). Because the organic intercalates are made by the reaction between the HgI_2 intercalate and alkylpyridinium iodide, there should be, if any, the same amount of unintercalated remnant of the pristine Bi2212 in both type of intercalates, resulting in similar magnetic behavior near the transition temperature (78 K) of the pristine material. However, HgI_2-Bi2212 and (Py-$C_nH_{2n+1})_2HgI_4$-Bi2212 exhibit different magnetic behavior at around 78 K (Figure 4A).

Furthermore, the organic intercalates show higher onset T_c values compared with the HgI_2 intercalate. Figure 4B shows the relation between T_c and the separation between the CuO_2 planes in adjacent blocks, where the T_c values of the intercalates are insensitive to the interlayer distance but are mainly dependent on the nature of the intercalant. Such findings are very interesting in light of the interlayer coupling theory in high-T_c superconductivity, which predicts that T_c is proportional to the coupling strength between adjacent superconducting layers, that is, inversely proportional to the layer separation [$k_B T_c \propto (\varepsilon d)^{-1}$, where k_B is the Boltzmann constant and e is the dielectric constant] [11]. The broadness of the superconducting transition observed for the organic intercalates (Figure 4A) is attributed to the thermal fluctuation in the true 2D superconductor [12].

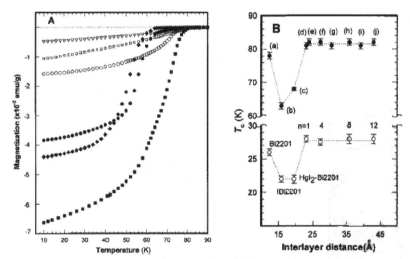

Figure 4. (A) Temperature dependence of ZFC magnetization with an applied magnetic field of 10 G, measured with a SQUID magnetometer. Data points represent Bi2212 (■), IBi2212 (◆), HgI_2-Bi2212 (●), and (Py-C_nH_{2n-1})2HgI$_4$-Bi2212 series [n = 1 (O), 4 (□), and 12 (5)]. For simplicity, data points for n = 2, 6, 8, and 10 are omitted. The shielding magnetization decreases sharply as the lattice expansion along the c axis increases. (B) The onset T_c values as a function of the interlayer distance of each sample [(a) Bi2212, (b) IBi2212, (c) HgI_2-Bi2212, and (d to j) the (Py-C_nH_{2n+1})2HgI$_4$-Bi2212 series with n = 1 (d), 2 (e), 4 (f), 6 (g), 8 (h), 10 (i), and 12 (j)], showing its insensitivity to interlayer distance. The open circles represent the Bi2201 series

The depressed T_c value upon HgI_2 intercalation is recovered by the intercalation of organic salt, which can be understood as a result of charge restoration of the host block. Because of the ionic-bonding character of the guest species (Py-C_nH_{2n+1})$_2$HgI$_4$ itself, a charge transfer between host block and intercalant layer would not be expected [13], in contrast to the halogen or mercuric halide intercalates.

To investigate the evolution of superconductivity in a single-layer superconductor, we also measured magnetizations for the organic intercalates of (Py-C_nH_{2n+1})$_2$HgI$_4$-Bi2201, where the basal increment (Δd) corresponds to 10.9, 13.1, 22.4, and 31.9 Å for n = 1, 4, 8, and 12, respectively. The superconductivity of Bi2201 is also retained upon intercalation of a long-chain organic compound. Moreover, although Tc is depressed ($\Delta Tc \approx 4K$) upon iodine and HgI_2 intercalation, all of the organic intercalates of Bi2201 show superconductivity with onset Tc values of 27 to 28 K, comparable to that of the pristine Bi2201 ($Tc \approx 26$ K) (Figure 4B). Such results allow us to conclude that Tc in the layered cuprate is essentially governed by the intrinsic property of a single CuO_2 plane rather than by the interlayer electronic coupling effect. Because there is only one CuO_2 plane per cuprate building block, the long-organic-chain–intercalated Bi2201 is believed to be a genuine single-layer superconductor because of the large interlayer distance compared with the c-axis coherence length of layered cuprates ($d \gg \xi c$). On the basis of these findings, it becomes clear that a 2D single cuprate sheet in the organic intercalate exhibits high-Tc superconductivity.

From the particle-size analyses and transmission electron microscopy (TEM) measurement, it is found that the plate-like superconducting colloidal particles are in the range of 50-200 nm (Figure 5). The electron diffraction (ED) pattern for colloidal particles exhibits characteristic pattern of the pristine Bi2212 lattice as shown in Fig. 2. All the diffraction patterns can be indexed as hl reflections for two-dimensional pseudotetragonal lattice of 5.4 × 5.4 Å, which is consistent with the host structure of Bi2212. According to the AFM height profiles [14], the vertical distance for the delaminated Bi2212 sheets was determined to be 20 Å, which is comparable to the thickness of unit building block of the pristine Bi2212. Such a finding can be regarded as an evidence of effective exfoliation of host lattice.

Figure 5. TEM and ED pattern of superconducting colloidal particles.

The film morphology plays an important role in determining the superconducting property such as critical current density (J_c). Especially, in order to attain a high-J_c, it is important to suppress the volume fraction and grain growth of second phases such as Bi-free phases and Cu-free phases, because such impurity phases can diminish a portion of superconductor and disturb the aligned grain growth of Bi-based superconductor. Therefore, the morphology of representative films was examined by using scanning electron microscopy (SEM). As shown in Figure 6(a), impurity phases could not be seen on the film heat-treated at

850 °C for 5 hrs. The most critical problem in achieving high-J_c film is the c-axis orientation, since the c-axis tilt and twist boundaries could act as weak links [15,16]. We found that the Bi2212 film is highly textured grain along c-axis in the cross-sectional image of Bi2212 film. Such a result indicates that delaminated Bi2212 nanosheets are expected to be excellent precursor for fabricating the superconducting thin or thick film and wire.

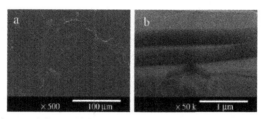

Figure 6. (a) The SEM surface morphology and (b) cross-sectional view of the fabricated superconducting film by EPD.

Bio-inorganic nanohybrids

The purpose of this study is not only to prepare new bio-inorganic nanohybrids but also to present biotechnological applications of inorganic materials, such as inorganic gene reservoirs or nonviral drug delivery carriers.

Nano-sized inorganic clay, layered double hydroxide (LDH), has been demonstrated as an excellent reservoir and delivery carrier for genes and drugs by hybridizing with DNA and antisense oligonucleotide (As-*myc*). According to X-ray diffraction pattern, the interlayer distance of LDH increases from 0.87 nm (for NO$_3^-$) to 2.39 nm (DNA), 1.94 nm (ATP), 1.88 nm (FITC), and 1.71 nm (As-myc), respectively, upon intercalating of biomolecules into hydroxide layers (Figure 7).

In the case of DNA-LDH hybrid, it was found that the hybrid has the gallery height of 19.1Å, which is consistent with the thickness of a DNA molecule (~20 Å) in a double helical conformation, with the interlayer DNA molecules arranged parallel to the basal plane of hydroxide layers. From the CD (cyclodichroism) analysis, it was determined that the intercalated DNA is stable between the hydroxide layers since the CD band of DNA-LDH hybrid was observed at the same wavelength compared with the band of ordinary B-form DNA.

In the case of DNA-LDH hybrid, it was found that the hybrid has the gallery height of 19.1Å, which is consistent with the thickness of a DNA molecule (~20 Å) in a double helical conformation, with the interlayer DNA molecules arranged parallel to the basal plane of hydroxide layers. From the CD (cyclodichroism) analysis, it was determined that the intercalated DNA is stable between the hydroxide layers since the CD band of DNA-LDH hybrid was observed at the same wavelength compared with the band of ordinary B-form DNA. Figure 8 represents the electrophoretic analysis of DNA-LDH hybrid, which shows that the DNA-LDH hybrid has pH dependent property. There are no DNA bands beyond pH $\cong$ 3, indicating that the DNA molecules in hybrid system are quite stable even in weak acidic atmosphere. However, the DNA bands appeared when the hybrids are treated in a strong acidic media below pH $\cong$ 2, since the hydroxide layers are dissolved in such acidic condition (lane 1-10). From the DNA elution as shown in lane 11 and 12, it can be deduced that the DNA-LDH hybrid can protect DNA from

DNase I enzyme. Consequently, the electrophoretic analysis reveals that the DNA-LDH hybrid plays a role as a gene reservoir.

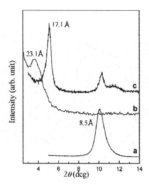

Figure 7. Powder X-ray diffractions for (a) the pristine LDH, (b) DNA-LDH, (c) As-myc-LDH.

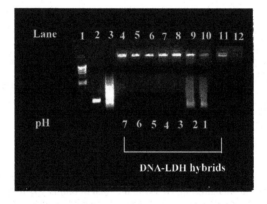

Figure 8. Electrophoresis analyses for the DNA–LDH hybrids with respect to pH. The pH of the solution dispersed with hybrid was adjusted to 7.5, 6.0, 5.0, 4.0, 3.0, 2.0 and 1.0, respectively, by adding 1M HCl. Lane 1; λ / Hind III cut DNA marker (descent to 23.1, 9.4, 6.5, 4.3, 2.3, 2.0 kbp), lane 2; 500 bp DNA marker, lane 3; DNA and lane 4-10; DNA–LDH hybrids at pH 7.5, 6, 5, 4, 3, 2, and 1, respectively. lane 11 ; DNA-LDH hybrid treated with DNase I and DNA recovered by acid treatment. Lane 12; DNA only treated with DNase I.

The cellular uptake experiment was carried out with As-*myc* LDH hybrids. We determined the abililty of As-*myc* LDH hybrids on the suppression of cancer cell division assuming that the biomolecule-LDH hybrid can be utilized as a delivery vector in gene therapy.

Figure 9 shows the effect of As-myc-LDH hybrid on the growth of cancer cell such as HL-60. The sequence of As-*myc* is 5' d (AACGTTGAGGGGCAT) 3', complementary to the initiation codon and the next four codons of c-*myc* mRNA, which can act as inhibitor for cancer cell. HL-60 cells treated with As-*myc*-LDH hybrids exhibit time-dependent inhibition on cell proliferation, indicating nearly 65 % of inhibition on the growth compared to the untreated cells, after 4 days. On the other hand, the growth of HL-60 cells treated with As-*myc* hybrid is only about 35 % compared with those treated with As-myc only.

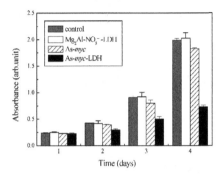

Figure 9. Effect of A-myc-LDH hybrids and As-myc only on the growth of HL-60 cells. Controlled cells are incubated without any treatment. The final concentration of each material was 20μM.

Since Figure 9 reveals that LDH itself does not inhibit the growth of HL-60 cells, the suppression effect of cancer cell is purely affected by As-myc-LDH. These results imply that As-*myc*s are incorporated into cells and eventually inhibit the growth of cancer cells thanks to the hybridization. It was also reported that the growth inhibition effect is time and dose dependent[11]. It is concluded that LDH can protect and deliver the intercalated oligonucleotide, and that interlayer As-myc can be effectively released from the hydroxide layer into cell fluids under physiological salt condition.

The other evidence on cellular uptake of the FITC-LDH hybrid was obtained directly from laser scanning confocal microscopy experiments. 1 μM and 3 μM FITC-LDH hybrids were added to NIH3T3 cells and then incubated for 1, 4, and 8 hrs. Then the cells were washed with PBS, and fixed with 3.7 % formaldehyde prior to measurements. Figure 10 shows the cellular localization of the fluorophore obtained after a fixed incubation time. The fluorophores were detected in cells within an hour of incubation, and the fluorescence intensities were increased continuously up to 8 hrs. The fluorophores in the cells were distributed mainly in peripheral and cytosol parts although some were found in the nucleus. Moreover, the cells treated with 3 μM of FITC-LDH hybrids showed more intense fluorescence than those with the 1 μM (Figure 10). In contrast, the cells treated with 5 μM FITC only remained dark regardless of incubation time, because cells could not take up FITC itself even with high concentration. It is obvious that LDH plays an important role in mediating the cellular uptake of FITC. All the cells can engulf the neutralized nanoparticles through phagocytosis or endocytosis. And then, it has been confirmed that FITC in the hybrid could be partially released by ion-exchange reaction in a physiological salt condition gradually over 8 hrs. Also, it has been found that LDHs as encapsulating materials could be gradually dissolved in a week acidic solution. Therefore, we conclude that the intracellular fluorescence has been created by deintercalated FITC and some by FITC-LDH hybrids in the cell.

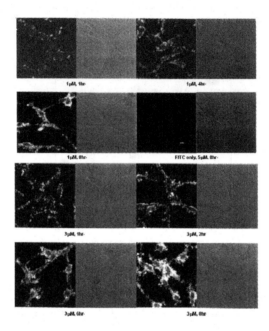

Figure 10. Laser confocal fluorescence microscopy of fluorophore in NIH3T3 cells. 6×10^4 cells/well were incubated with (a)1 mM FITC-LDH for 1, 4, and 8 hrs respectively (The other fluorescence microphotograph was obtained with 5 mM FITC only), (b) 3 mM FITC-LDH for 1, 2, 6, and 8 hrs respectively. The bar is 10 mm.

CONCLUSION

New classes of inorganic-inorganic heterostructures with high photocatalytic activities can be realized via exfoliation-reassembling method. And organic-inorganic heterostructures with high-T_c superconductivity can be synthesized by hybridizing metal halides or organic salts with Bi-based cuprates. These superconducting compounds are believed to be promising precursor materials for superconducting nanoparticles, thin or thick films, and wires. And also, we are able to demonstrate that the biomolecules can be intercalated into LDH via ion-exchange reaction to construct bio-inorganic nanohybrid and that inorganic supramolecules, such as the LDHs with nanometer size, can play excellent roles as reservoir for biomolecules and as delivery carrier for gene and drugs.

ACKNOWLEDGEMENTS

This work was supported by the SRC program of the Korea Science and Engineering Foundation (KOSEF) through the Center for Intelligent Nano-Bio Materials at Ewha Womans University (grant: R11-2005-008-01001-0).

REFERENCES

[1] J. H. Choy, N. G. Park, S. J. Hwang, D. H. Kim, and N. H. Hur, *J. Am. Chem. Soc.*, **116**, 11564 (1994).

[2] J. H. Choy, S. J. Kwon, and G. S. Park, *Science*, **280**, 1589 (1998).

[3] J. H. Choy, S. Y. Kwak, J. S. Park, Y. J. Jeong, and J. Portier, *J. Am. Chem. Soc.*, **121**, 1399 (1999).

[4] J. H. Choy, H. C. Lee, H. Jung, H. Kim, and H. Boo, *Chem. Mater.*, **14**, 2486 (2002).

[5] T. Sasaki, M. Watanabe, H. Hashizume, H. Yamada and H. Nakazawa, *J. Am. Chem. Soc.*, **118**, 8329 (1996).

[6] S. M. Paek, H. Jung, M. Park, J. K. Lee, and J. H. Choy, *Chem. Mater.*, **17**, 3492 (2005).

[7] E. Scolan and C. Sanchez, *Chem. Mater.*, **10**, 3217 (1998).

[8] J. K. Lee, W. Lee, T. J. Yoon, G. S. Park, and J. H. Choy, *J. Mater. Chem.*, **12**, 614 (2002).

[9] J. H. Choy, S. J. Kwon, S. H. Hwang, Y. I. Kim, and W. Lee, *J. Mater. Chem.*, **9**, 129 (1999).

[10] X. D. Xiang, S. Mckernan, W. A. Vareka, A. Zettl, J. L. Corkill, T. W. Barbee, and M. L. Cohen, *Nature*, **348**, 145 (1990).

[11] D. R. Harshman, and A. P. Mills, *Phys. Rev. B*, **45**, 10684 (1992).

[12] P. A. Bancel, and K. E. Gray, *Phys. Rev. Lett*, **46**, 148 (1981).

[13] H. Selig and L. B. Ebert, *Adv. Inorg. Chem. Radiochem.*, **23**, 281 (1980).

[14] J. H. Choy, S. J. Kwon, S. H. Hwang, and E. S. Jang, *Mater. Res. Soc. Bull.*, **25**, 32 (2000).

[15] L. N. Bulaevskii, L. L. Daemen, M. P. Maley, and J. Y. Coulter, *Phys. Rev. B.*, **48**, 13798 (1993).

[16] B. Hensel, G. Grasso, and R. Flükiger, *Phys. Rev. B.*, 51, 15456 (1995).

SINGLE-CRYSTAL SIC NANOTUBES: MOLECULAR-DYNAMIC MODELING OF STRUCTURE AND THERMAL BEHAVIOR

V.L. Bekenev[1], V.V. Kartuzov[1], Y. Gogotsi[2]

[1] I.N. Frantsevich Institute of Problems of Materials Science NASU, Kiev, Ukraine
[2] A.J. Drexel Nanotechnology Institute and Department of Materials Science and Engineering, Drexel University, Philadelphia, PA 19104, USA

This effort is to evaluate a new type of nanostructures – polyhedral single-crystal SiC nanotubes. Articles [1, 2] discussed the synthesis of nanotubular forms of silicon carbide. The nanotubular SiC in these works was produced by a chemical reaction between SiO vapor and multil-walled carbon nanotubes. Investigations of the samples obtained via this method have not revealed their crystal structure. However, the X-ray diffraction studies [1] showed that tubes with the average internal diameter 900 Å consist of cubic silicon carbide. Work [3] suggested that nanotubes may be single crystals; in fact, it is probably the first time the terminology "a single crystal nanotube" was used. In [3], the nanotubes of gallium nitride with internal diameter 300-2000 Å and external diameter 350-2500 Å were obtained by the method of epitaxial casting. X-ray and electron diffraction measurements show that the synthesized tubes are single crystal and have würtzite structure. The axis of tube coincides with a [0001] crystal direction. It is important to note that nanotubes from würtzite gallium nitride radically differ from all other inorganic nanotubes (BN, NiCl$_2$, MoS$_2$, VO$_x$) in that they are not layered or disordered structures. The atoms between the external and internal surfaces of the tube possess a regular crystal arrangement. Based on this fact, the work presented here suggests a model of single crystal tubes for crystals with cubic (diamond, cubic silicon carbide, boron nitride, gallium nitride) and hexagonal diamond structures (lonsdalite, würtzite silicon carbide, boron nitride, gallium nitride).

One may evaluate a diamond crystal structure as hexagonal structure if a direction [111] is chosen as axis c. Thus the parameters of a hexagonal cell of crystal are: $a_g = \dfrac{a_c}{\sqrt{2}}$, $c_g = a_c\sqrt{3}$, where a_c is a parameter of a cubic cell of crystal. Let's "cut off" a regular hexagon prism from a crystal whose axis coincides with a direction [111], and a side of the hexagon is ma_g, $m > 1$. From the obtained prism let's once again "cut off" a similar prism of smaller size with a side of hexagon na_g, $0 < n < m$. After these operations we shall obtain a hexagonal prismatic nanotube. Inside the tube's body the atoms are arranged as in the initial crystal. At $n = 0$ in lieu of a tube we would have a hexagonal single crystal nanorod or nanowire. For a designation of single crystal tubes we would apply a symbol {m, n}. For the construction of nanotubes from würtzite-like crystals as tube axis we pick up the direction [0001] and similar operations are carried out.

Molecular-dynamic modeling of single crystal nanotubes was carried out with the use of well-known Tersoff's potentials [4]. We consider a case of infinite length tubes realized by imposing periodic boundary conditions along the tube axis. A length of periodically repeated part of tube is chosen as $5c_g$. In this case, when equilibrium configuration is attained, energy changes were less than 10^{-4} eV/atom for two consecutive time steps. Temperature stability of nanotubes was investigated under a condition of volume constancy. Single crystal tubes {m, n}, m=3-6, n=1, 2, ..., (m-1) were evaluated. Figures 1, 2 illustrate the results of calculations.

While it is known that surface reconstruction or tube surface termination with hydrogen or oxygen may affect the total energy of the nanotube, they were not taken into consideration in this preliminary study.

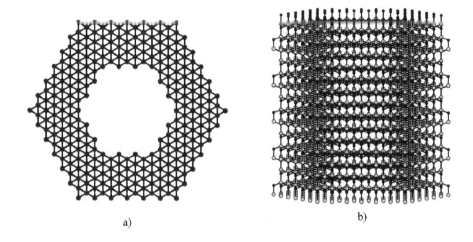

Fig. 1. Single crystal 3C-SiC {6,3} nanotube: a) cross-section; b) side view. Dark circles – carbon atoms, light – silicon atoms.

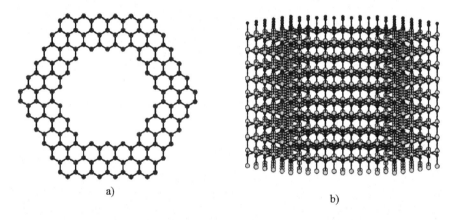

Fig. 2. Single crystal 2H-SiC {6,3} nanotube: a) cross-section; b) side view. Dark circles – carbon atoms, light – silicon atoms.

Energy of formation of a single crystal nanotube is defined as the energy difference per one atom in nanotube and corresponding crystal. Dependence of energy formation on wall thickness is presented in Fig. 3. For thin tubes ({3,1}, {4,2}, {5,3}, {6,4}) energy of formation essentially exceeds atomic energy in the corresponding crystal (Fig. 3). As tube thickness increases, the energy of formation decreases and becomes comparable with energy of deformation for carbon nanotubes of small diameter. One may expect that it would be possible to synthesize tubes with (m–n) > 4. For GaN nanotubes, reported in [3], a value of difference is m–n = 16–160. From Fig. 3 we also see that equilibrium thickness of tubes weakly depends on m, n and is defined by the difference (m–n).

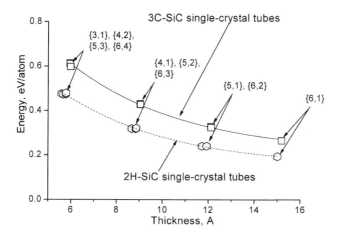

Fig. 3. Energy of formation of single crystal nanotube as a function of tube wall thickness.

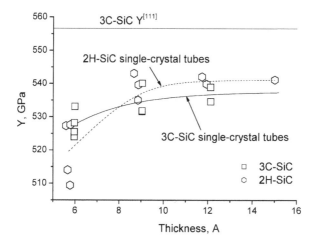

Fig. 4. Young's moduli of single crystal SiC nanotubes with various wall thicknesses.

Dependence of Young's modulus, which is a good indicator of mechanical properties, on thickness of tube walls is rather insignificant, as can be seen from fig. 4. In the same figure a horizontal line corresponds to Young's modulus of cubic silicon carbide in a [111] direction. Young's moduli of single crystal tubes are lower by ~3-7 % than theoretical values of Young's modulus for 3C-SiC.

Thermal stability of single crystal tubes was also investigated. For this purpose caloric curves were calculated, i.e. dependences of cohesive energy of tubes on kinetic temperature, as seen in Fig. 5. For both materials caloric curves are divided into the following categories: a solid state, a transition from a solid state to liquid, a liquid, transition from liquid to vapor, and vapor.

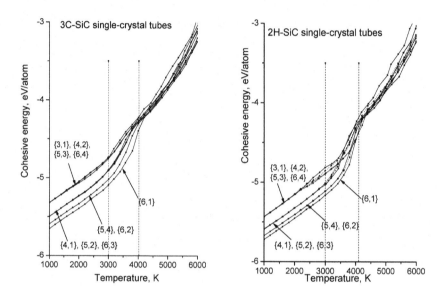

Fig. 5. Cohesive energy of single crystal nanotubes as a function of temperature.

On the obtained curves for single crystal tubes one may allocate three areas. In the first area a dependence of cohesive energy on temperature possesses a linear character. In this area the ordered structure of the tube is kept. The second area begins with a nonlinear growth of cohesive energy associated with the beginning of collapse of a tube. Temperature at the beginning of nonlinear growth of cohesive energy is approximately identical for tubes of various thicknesses. Figs. 6, 7 illustrate cross-sections of nanotubes at various temperatures. Collapse of tube structure occurs by "evaporation" of complexes of atoms or separate atoms in the internal area of the tube. The third area on the curves corresponds to fast degradation process with the tube completely collapsing.

Finally, if produced in large quantities, SiC nanotubes may find many applications due to their useful properties. They may compete with SiC whiskers and fibers [5] as a light-weight reinforcement for metal and ceramic matrix composites. SiC is a wide gap semiconductor having a much higher temperature stability than Si. Thus, single-crystal nanotubes may be used in nanoscale electronic devices.

SiC nanotubes will have a much high oxidation resistance in air compared to carbon or boron nitride nanotubes. Of course, many other applications in which carbon nanotubes or Si nanowires are currently used may be open for SiC nanotubes as well.

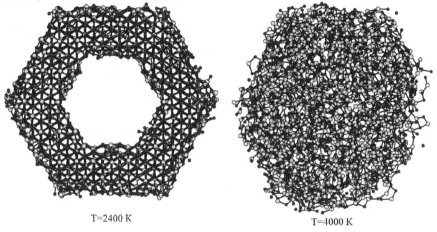

T=2400 K   T=4000 K

Fig. 6. Cross-section of a 3C-SiC {6,3} nanotube at different temperatures.

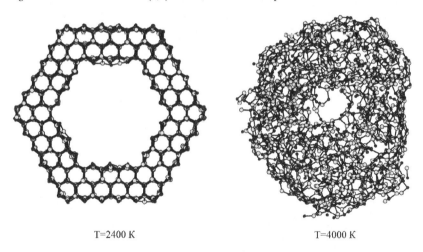

T=2400 K   T=4000 K

Fig. 7. Cross-section of a 2H-SiC {6,3} nanotube at different temperatures.

SUMMARY

A model of single crystal nanotubes having sphalerite-like and würtzite-like crystal structures has been developed. The investigation of mechanical properties by the method of molecular dynamics modeling with the use of Tersoff's potentials showed that Young's moduli of all considered nanotubes possess smaller values than those of carbon and BN nanotubes formed by curtailing of graphite or boron nitride layers. , Analysis of the temperature dependence of cohesive energy of single crystal 3C and 2H SiC nanotubes show that they remain stable up to temperatures ~3000 K, and the temperature of the beginning of tube collapse depends only weakly on wall thickness. Energy of formation of single crystal {m, n} 3C and 2H SiC nanotubes at (m-n)> 4 of ~0.2 eV/atom allows us to believe in possibility of synthesis of these tubes.

Acknowledgments: This effort was financed in part by NAS of Ukraine under the "Nanocrystalline materials" program.

REFERENCES

1. C. Pham-Huu, N. Keller, G. Ehret and M. J. Ledoux, The first preparation of silicon carbide nanotubes by the shape memory synthesis and their catalytic potential, *J. Catal.* 2001. **200** (2) pp. 400-410.
2. X.-H. Sun, C.-P. Li, W.-K. Wong, N.-B. Wong, C.-S. Lee, S.-T. Lee, B.-K. Teo, Formation of silicon carbide nanotubes and nanowires via reaction of silicon (from disproportionation of silicon monoxide) with carbon nanotubes, *J. Am. Chem. Soc.* 2002. **124** (48) pp. 14464-14471.
3. J. Goldberger, R. He, Y. Zhang, S. Lee, H. Yan, H.-J. Choi, P. Yang, Single-crystal gallium nitride nanotubes. *Nature.* 2003. **422**. pp. 599-602.
4. J. Tersoff. Modeling solid-state chemistry, Interatomic potentials for multicomponent systems, *Phys. Rev.* 1989. **B39** (8) pp. 5566-5568. Erratum: *Phys. Rev.* 1990. **B41** (5) p. 3248.
5. K.L. Vyshnyakova, L.N. Pereselentseva, Z.G. Cambaz, G.N. Yushin, Y. Gogotsi, Whiskerisation of polycrystalline SiC fibres during synthesis, *British Ceramic Transactions,* 2004. **103** (5) pp. 193-196

VIBRATIONAL SPECTRUM OF A DIAMOND-LIKE FILM ON SiC SUBSTRATE

V. Shevchenko*, Y. Gogotsi**, E. Kartuzov*

* Frantsevich institute of Problems in Materials Science, National Academy of Science of Ukraine, Kiev , Ukraine
** Department of Materials Science and Engineering and A.J. Drexel Nanotechnology Institute, Drexel University, Philadelphia, PA 19104, USA

Vibrational spectra of diamond-like film grown on the (111) surface of 3C-SiC were calculated by a method of molecular dynamics. A certain similarity in the density of vibrational states simulated for a free surface of diamond and the diamond-like film on SiC surface is observed.

INTRODUCTION

Formation of various carbon films on a SiC surface has been investigated experimentally [1] and by computer simulations. Modeling of the films was carried out by the method of molecular dynamics with use of Tersoff's potentials for the Si-C system [2]. The authors of work [1] have noted a potential opportunity for the formation of thin diamond-containing films on the (111) SiC surface. SiC is also used as a substrate for diamond growth. Structural investigations of diamond films on SiC have been performed. However, a comparison of inelastic dissipation of electrons by these films has not been carried out, as phonon spectra of these objects were not investigated. Therefore in this effort the vibrational spectra of diamond-like films covering the (111) surface of SiC have been calculated. We picked up for our analysis the atomic structure obtained in [1].

METHODS OF CALCULATION

A calculation of vibrational spectra in this work is based on a method offered in Ref. [3]. According to [3], initially a speed - speed autocorrelation function (SACF) is calculated for some groups of atoms, and then the Fourier transformation of the found function is applied. In comparison with a standard method, in which the frequencies of the appropriate dynamic matrix are determined, the method based on the application of autocorrelation function possesses an advantage that allows us to account for a nonlinearity of interatomic interaction.

Instantaneous speeds of the selected group of atoms required for a calculation of appropriate SACF are recorded (with a small time interval between the records) in a process of molecular dynamics. Temperature of the system, a time interval between the consecutive records of speeds, and a total number of these records are the important parameters which define the quality of the appropriate autocorrelation function and, thus, the required vibrational spectrum. Let's note that the described method allows us to calculate the local frequency characteristics of an oscillating system and thus is convenient to solve the problems formulated in this work.

SACF $Z(t)$ is calculated by the following formula [3]

$$Z(t) = \sum_{\tau}^{\tau_m} \frac{\langle \vec{v}_n(t+\tau) \cdot v_n(\tau) \rangle}{\langle \vec{v}_n(t) \cdot v_n(\tau) \rangle},$$

where the angle brackets specify the averaging for all atoms of the selected group of atoms. To obtain the spectrum $Z(\omega)$ associated with the selected atoms, the Fourier transformation [4] is performed

$$Z(\omega) = \int dt e^{i\omega t} Z(t).$$

Oscillations of the investigated system were modeled by a common method of molecular dynamics. The interaction among atoms was described by known empirical Tersoff's potentials.

RESULTS AND DISCUSSION

Preliminary calculations enabled us to choose optimum parameters, namely, temperature of the system – 500 K, integration step of the equations of movement (molecular dynamics) – 10^{-16} sec, the number of consecutive records of atoms speed – 1000, a temporary interval between the records of speeds – 3.0×10^{-16} sec, and a total amount of atoms of cluster – 1000.

With the aim to test the selected method we calculated the vibrational spectrum of diamond. Fig. 1 illustrates the obtained SACF and phonon spectrum of diamond.

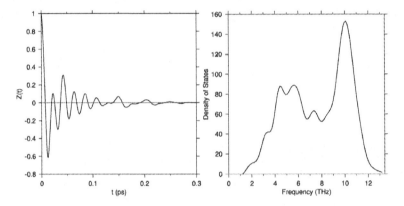

Fig. 1 SACF $Z(t)$ and phonon spectrum of a diamond cluster

A deviation of the calculated spectrum from experimental data may be explained by insufficient accuracy of Tersoff's parameters and rather small sizes of the super cell of modeling.

In a similar way, we calculated local densities of vibrational states which correspond to:
• the atoms located on and nearby free (relaxed) surface (111) of diamond, and
• the atoms of the diamond-structured film adsorbed on the 3C-SiC surface (111).
The obtained phonon densities are presented in Fig. 2.

It is clear that the density of vibrational states corresponding to a diamond-structured film on the SiC surface are in general similar to the densities of phonon oscillations corresponding to a free surface of diamond. It may serve as additional evidence that carbon films obtained by modeling are really diamond-structured. A shift of the peaks corresponding to diamond-structured film to the left relatively similar peaks, corresponding to the surface of diamond, is caused by the fact that the diamond-structured film is slightly stretched. A change of the peak's shape and intensity is stipulated by the influence of the substrate material (SiC) on vibrational properties of the diamond film.

Certainly, a direct experimental investigation of phonon spectra of carbon films on SiC surface will be required to answer some of the questions raised in this work. For example, such question is the investigation of non-elastic dissipation of electrons by carbon films on the SiC substrate. We are not aware of any published studies on that topic.

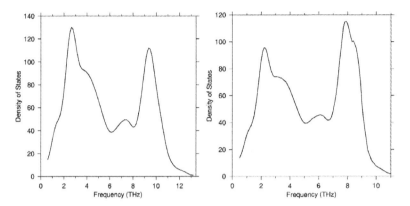

Fig. 2 Densities of phonon states corresponding to surfaces of diamond (left) and a diamond-structured film on a SiC substrate (right)

CONCLUSIONS

The certain similarity of density of states calculated for the surface of diamond and a diamond-structured film on the (111) SiC surface is observed. This observation suggests the possibility of direct growth of diamond and diamond-like films on the silicon carbide surface.

REFERENCES

1. Y. Gogotsi, V. Kamyshenko, V. Shevchenko, et al. Nanostructured carbon coatings on silicon carbide: experimental and theoretical study, *Proceeding NATO ASI on Functional Gradient Materials and Surface Layers Prepared by Fine Particle Technology*, Ed. by M. I. Baraton, I. Uvarova. Dordrecht, NL: Kluwer, 2001, 239-255.
2. J. Tersoff, Modeling solid-state chemistry: Interatomic potentials for multi-component systems, *Phys. Rev. B.* 1989, **39**(8) 5566-5568.
3. E. Kim and Y. H. Lee, Structural, electronic, and vibration properties of liquid and amorphous silicon: Tight-binding molecular-dynamics approach, *Phys. Rev. B.* 1994 **49**(3) 1743 -1749.
4. G. M. Jenkins and D. G. Watts, Spectral Analysis and its Applications. Holden-Day, 1969.

Author Index